GROVER E. MURRAY
STUDIES IN THE
AMERICAN SOUTHWEST

Also in this series:

Agaves, Yuccas, and Their Kin: Seven Genera of the Southwest, by Jon L. Hawker

Artists of the Horse: Capturing the Spirit of America's West, Heidi Brady and Scott White

Auto Touring America's National Parks: The Photography of H. A. Spallholz, by Julian E. Spallholz, Lance O. Spallholz, and Arthur S. Vaughan

Between Two Rivers: Photographs and Poems Between the Brazos and the Rio Grande, by Jerod Foster and John Poch

Birds of the Texas Coast: Photographs, by B. C. Robison, photographs by Ron Grimes

Blackdom, New Mexico: The Significance of the Afro-Frontier, 1900–1930, by Timothy E. Nelson

Brujerías: Stories of Witchcraft and the Supernatural in the American Southwest and Beyond, by Nasario García

Cacti of Texas: A Field Guide, by A. Michael Powell, James F. Weedin, and Shirley A. Powell

Cacti of the Trans-Pecos and Adjacent Areas, by A. Michael Powell and James F. Weedin

Cowboy Park: Steer-Roping Contests on the Border, by John O. Baxter

Dance All Night: Those Other Southwestern Swing Bands, Past and Present, by Jean A. Boyd

Dancin' in Anson: A History of the Texas Cowboys' Christmas Ball, by Paul H. Carlson

Deep Time and the Texas High Plains: History and Geology, by Paul H. Carlson

"Don't Count the Tortillas": The Art of Texas Mexican Cooking, by Adán Medrano

Down in the Dirt: Graphic Art by Terry Allen, 1970–2025, by Peter S Briggs

The Edge Rover: The Life and Times of Mountain Man Isaac Slover, by Timothy E. Green

Engineering the Modern American West: The Life of Frederick Haynes Newell, by Cole Bedford

Equal Opportunity Hero: T. J. Patterson's Service to West Texas, by Phil Price

Finding the Great Western Trail, by Sylvia Gann Mahoney

From Texas to San Diego in 1851: The Overland Journal of Dr. S. W. Woodhouse, Surgeon-Naturalist of the Sitgreaves Expedition, edited by Andrew Wallace and Richard H. Hevly

The Frontier Centennial: Fort Worth and the New West, by Jacob W. Olmstead

Grasses of South Texas: A Guide to Identification and Value, by James H. Everitt, D. Lynn Drawe, Christopher R. Little, and Robert I. Lonard

A Haven in the Sun: Five Stories of Bird Life and Its Future on the Texas Coast, by B. C. Robison

A Kineño's Journey: On Learning, Family, and Public Service, by Lauro F. Cavazos, with Gene B. Preuss

Kit Carson and the First Battle of Adobe Walls: A Tale of Two Journeys, by Alvin R. Lynn

In the Shadow of the Carmens: Afield with a Naturalist in the Northern Mexican Mountains, by Bonnie Reynolds McKinney

Javelinas: Collared Peccaries of the Southwest, by Jane Manaster

Land of Enchantment Wildflowers: A Guide to the Plants of New Mexico, by LaShara J. Nieland and Willa F. Finley

Little Big Bend: Common, Uncommon, and Rare Plants of Big Bend National Park, by Roy Morey

Lone Star Wildflowers: A Guide to Texas Flowering Plants, by LaShara J. Nieland and Willa F. Finley

More Than Running Cattle: The Mallet Ranch of the South Plains, by M. Scott Sosebee, photographs by Wyman Meinzer

My Wild Life: A Memoir of Adventures within America's National Parks, by Roland H. Wauer

Myth, Memory, and Massacre: The Pease River Capture of Cynthia Ann Parker, by Paul H. Carlson and Tom Crum

Opus in Brick and Stone: The Architectural and Planning Heritage of Texas Tech University, by Brian H. Griggs

Pecans: The Story in a Nutshell, by Jane Manaster

Picturing a Different West: Vision, Illustration, and the Tradition of Austin and Cather, by Janis P. Stout

Plants of Central Texas Wetlands, by Scott B. Fleenor and Stephen Welton Taber

Seat of Empire: The Embattled Birth of Austin, Texas, by Jeffrey Stuart Kerr

Texas Natural History in the 21st Century, by David J. Schmidly, Robert D. Bradley, and Lisa C. Bradley

Texas, New Mexico, and the Compromise of 1850: Boundary Dispute and Sectional Crisis, by Mark J. Stegmaier

Texas Quilts and Quilters: A Lone Star Legacy, by Marcia Kaylakie with Janice Whittington

Truly Texas Mexican: A Native Culinary Heritage in Recipes, by Adán Medrano

The Wineslinger Chronicles: Texas on the Vine, by Russell D. Kane

THE STORY THE BONES TELL

A HISTORY OF THE LUBBOCK LAKE LANDMARK

SALLY LOGUE POST

FOREWORD BY EILEEN JOHNSON

TEXAS TECH UNIVERSITY PRESS

This book is typeset in Adobe Caslon Pro. The paper used in this book meets the minimum requirements of ANSI/NISO Z39.48-1992 (R1997). ♾

Designed by Hannah Gaskamp
Cover design by Hannah Gaskamp
Cover image: Vance Holliday standing in front of the strata at the Lubbock Lake Landmark (courtesy of Lubbock Lake Landmark)

Library of Congress Cataloging-in-Publication Data on file

ISBN: 978-1-68283-331-5 (paperback)
ISBN: 978-1-68283-332-2 (ebook)

Texas Tech University Press
Box 41037
Lubbock, Texas 79409-1037 USA
800.832.4042
ttup@ttu.edu
www.ttupress.org

For Dr. Eileen Johnson

CONTENTS

ILLUSTRATIONS

FOREWORD

This venture actually started many, many years ago when Sally Post was working for University News and Publications with Bea Zeeck. Bea was an early supporter of the Lubbock Lake Landmark. Through the years, Sally wrote more stories on both the Landmark and the Museum of Texas Tech University (now Museum at Texas Tech University). Sally became my go-to person, not only for Landmark publicity but also for navigating the inner workings of campus administration. Sally's involvement with the Landmark grew over the years, particularly after she retired.

The fiftieth anniversary of field research and public education at the Landmark marked a watershed moment—a time to look back on all that has been accomplished and a time to look forward to the many Landmark opportunities still presenting themselves in research and education. This book represents one of those opportunities. Public education and involvement with the Landmark are high priorities that go hand in hand with the research. A hallmark of the Landmark's public programs is that they are based on the Landmark's research findings and within its regional research program. The programs translate those scientific findings and interpretations for the Landmark's various audiences. Sally's book is another translational piece based on scientific findings and interpretations, but couched in terms for a general audience.

As Sally has found out, the Landmark is many things underlain by complexity and nuances, as well as the political framework that makes the Landmark's protection, research, and public programming possible. Furthermore, the Landmark does not readily reveal its secrets. Sally made a valiant effort to unravel and understand the numerous concepts and principles from various disciplines interwoven into the interdisciplinary research at the Landmark. I am a Quaternary research scientist and an eclectic one at that. The entire Quaternary is fair game, and that perspective comes through in the regional research program. Archaeology, having to do with past peoples and lifeways, is one of those Quaternary sciences and comes into play beginning in the late Pleistocene. Sally has endeavored to understand and relate in her book the fuller range of those sciences employed in the Landmark's research program. Equally challenging was her exploration of my perspective and how it has evolved and grown over the years.

For me, this venture has been a continual and enjoyable blast from the past—at times a stroll and other times a sprint down memory lane. It has brought reflection, a consideration of the past, and an approach forward in the next chapter of Landmark research and public education.

EILEEN JOHNSON, PHD
DIRECTOR, LUBBOCK LAKE LANDMARK

PREFACE

A Different Type of Dig

Sometimes archaeologists know where to dig by happenstance—perhaps because someone saw bones eroding from the side of a draw. Recovering those bones, however, is rarely straightforward. Time and gravity change everything, requiring persistence, if not urgency, and ingenuity—even if the latter means hanging from a ladder to recover bones from atop a fragile column that is near collapse.

The idea for this book began with a passing remark I made to Lubbock Lake Landmark Director Eileen Johnson. I told her it was surprising that no book about the Landmark's history and significance existed for a general audience. After all, it is one of the most noteworthy archaeological sites in the Americas. She looked at me for a moment and said, "Then you should write one."

I remember thinking, *OK, I know about the Landmark. I have interviewed Dr. Johnson numerous times in my years at Texas Tech University and spent time at the site. How hard could it be?*

Silly me. It turns out that writing this book was bloody hard, but it has also been fascinating and fun. My initial idea was to tell the stories that the bones reveal. I have ended up with a story of research, education, public outreach, and the political and financial struggles to maintain the Landmark and continue uncovering its secrets.

So, in 2022, I began a dig of my own, though it was very different from the ones pursued in the hot West Texas sun. I devoted hours to conducting interviews, digging through the Landmark's archive, going through boxes of records, and listening to oral histories in Texas Tech University's Southwest Collection/Special Collections Library.

I once told a friend that my dream job would be spending all day immersed in the Southwest Collection's historical documents. I love delving into the rich history and stories it holds. And indeed, the hours spent reading the material were enjoyable. But the reality of the task soon set in as I dealt with tired eyes, paper cuts, and allergy attacks.

The fun part was following the information from one box of materials to another. I expended hours going through a lot of boxes. These papers and interviews sent me down myriad pathways, serially chasing another bit of information.

I also spent hours reviewing archival material housed in the Landmark offices at the Nash Interpretive Center and poring over old newsletters. I found a story about what researchers initially thought was the discovery of a camel's jawbone. However, subsequent research revealed that the creature in question was a rhinoceros. For me, the more remarkable story here is not only the work it takes to get the identification of a bone right but also the fact that thousands to millions of years ago, camels, rhinos, Columbian mammoths, and fearsome ancient lions could have walked across what is now the Texas Tech University campus.

Landmark volunteer manuals and educational materials were rich with information. The more materials I sorted through from every source, the more interviews I needed to conduct—and the more I realized that researching a book is somewhat akin to an archaeological dig.

Indeed, it proved a painstaking excavation, discovery, and verification process. Following the paper trail, I discovered much more than the archaeological side of the landmark. The Landmark crews unearth bones and artifacts that give us glimpses into life on the Llano Estacado dating back thousands of years. The archives and my interviews also provided a much broader story highlighting the various coalitions that Curry Holden, Bob Nash, John Montford, and Dr. Johnson wove together to ensure the early survival of the Landmark and the struggles it faces as it continues discovering the stories of this magical place.

There were also frustrations along the way. I assumed (yes, I know, one should never do that) there would be abundant documents referencing a critical source, the late Bob Nash, who played a significant role in the Landmark's progress. While there was material on his radio and television career, I found little information about his involvement with the Landmark. Thus, the hunt began to find people who had known Nash and remembered his love of history in general and of the Landmark in particular. I found a few people who could share stories of Bob's passion for the Landmark; it is fortunate, however, that Dr. Johnson had documented his involvement in a book chapter she wrote in 2016.

All my working life, I've made my living by writing. Yet, this project differs from any other I have undertaken. Ultimately, I'm glad I became the person Dr. Johnson trusted to tell the Landmark's story. I hope you will be as excited as I was to find that horses first evolved in North America, as did camels and other unexpected creatures. Perhaps the most fascinating fact I discovered is that humans have called Lubbock home for at least 13,000 years.

AUTHOR'S NOTE

I wrote this book for a general audience. It is partly the story of the bones of animals, some as old as millions of years, and the human history of those who have called the Southern High Plains home for at least 13,000 years. I've also tried to tell the story of the people who have preserved and protected the Lubbock Lake Landmark, one of the most significant archaeological sites in the Americas.

Over the years, people have used various names to refer to what is now formally known as the Lubbock Lake National Historic Landmark. In 1977, the National Park Service so designated it, and in 1981, the Texas Historical Commission named it a State Archeological Landmark. Most people now refer to the site as the Lubbock Lake Landmark, or simply the Landmark. For simplicity, I use these latter two names, even when discussing its earliest history.

Over approximately one hundred hours from September 2022 to October 2024, I interviewed forty-one individuals in person or via Zoom, telephone, or email. Each shared a story about what the Landmark has meant to them personally and professionally, detailing the efforts to preserve and grow the site's research and education efforts. Most of the information in the book comes from these interviews. Additional sources are cited in the endnotes.

Two booklets from the Lubbock Lake Landmark Natural History Series provided extensive information on the Landmark's cultural history and the ancient animals that lived around the area. Eileen Johnson and Susan Shore published *Mammals at Lubbock Lake Landmark* in 2010. Mindy Myers, Eileen Johnson, Susan Shore, Vance T. Holliday, and Susan Rowe published *The Lubbock Lake Landmark: An Archaeological and Natural History Preserve* in 2012. (While this booklet is out of publication as of 2026, work on a new edition is underway.) Handbooks also provide essential details, but they are reserved for volunteers.

Next, I have chosen to use calendar year dates rather than radiocarbon dates in this book. Radiocarbon dating (^{14}C) is the most accurate method for determining the age of organic materials. It is also the date most scientists use, including those at the Lubbock Lake Landmark. However, calendar dates have become more widely accepted in academic and general publications because they are more familiar to the public. There is a difference between the two types of measures. A 13,000-calendar-year sample of Clovis-age material will yield a ^{14}C date of 11,000 years. The difference occurs when a computer program translates the ^{14}C date into calendar years, somewhat like the difference that arises when meters are converted to feet. Radiocarbon dating is the foundation of the computer programs that calibrate ^{14}C years to calendar years. I note when a radiocarbon date is used.

Lastly, scientific methods form the basis of everything researchers do at the Landmark. However, I have tried to keep the science simple and focus on the people and ancient creatures that have called the South Plains region home.

THE STORY THE BONES TELL

CHAPTER 1

IT'S ALL ABOUT THE WATER

A 13,800-year-old wolf lies next to a Columbian mammoth on a fragile eight-foot-by-four-foot column of dirt. The Lubbock Lake Landmark excavation crew, which found the bones in 2019, realizes such a pairing is almost unheard of. The discovery fuels a dramatic race against the clock. Every scrape of the trowel sends vibrations through the column, already eroded loose from the cliff face.

Balancing on a ladder in the 100-degree heat, John Moretti, then the senior crew chief at the Landmark, works inch by inch, scraping away the Ice Age sediment around the bones. After about an hour, he uncovers two more surprises: an intact ancient horse pelvis and half of an extinct box turtle shell, both of which are of a similar age.

Two days later, the column collapses, but not before the team collects many of the mammoth and wolf bones, along with the horse pelvis. As with any excavation, the fossils are logged and photographed in situ, providing scientific context.

Figure 1. John Moretti, then a research aide and senior crew chief, climbed one of two ladders to perform a tricky excavation. (Courtesy of Lubbock Lake Landmark; Stance Hurst, photographer)

While odd, Moretti says, finding the two creatures together is not a mystery. The deposits demonstrate that Spring Creek flowed through the area at one time. "The water most likely acted as a transport and concentrated the remains from across the area to one place," he says.

Although the team followed safety precautions and paleontological excavation protocols, the way this excavation played out is substantially different from a standard archaeological dig. Crews typically work methodically at ground level, digging in one-meter-by-one-meter squares.

While working on a ladder makes for an extraordinary excavation, the fossils themselves are not rare finds for the Landmark crews. Excavations at the Landmark and across the region have unearthed a menagerie of Ice Age animals, ranging from the rare American lion to Columbian mammoths, short-faced bears, ancient camels, and horses.

Driving the decades-long research program is Dr. Eileen Johnson, the Landmark's director. While she doesn't work in the field as often as she once did, she understands that the thrill of discovery brings men and women from around the world to volunteer to work during the scorching South Plains summers. She explains, "What I remember most about field seasons is the anticipation and excitement of discovery, that with the next brush stroke or pass of the trowel, something unexpected would appear. The Lubbock Lake record is full of surprises."

It is almost impossible to talk about the Lubbock Lake Landmark without also talking about Johnson. She followed her mentor, Craig Black, to Texas Tech University in 1972, intending to stay about three years and complete her doctorate. More than fifty years later, she continues to lead the research program and has earned international recognition as a researcher and respected professor.

The Lubbock Lake Landmark sits on the northern edge of the City of Lubbock, in a U-shaped bend of Yellowhouse Draw. It spans approximately 336 acres, about five percent of which has been explored by excavation. The Landmark is the hub of about half a dozen other excavation sites spread across the South Plains. The broader regional research helps frame the Landmark's discoveries within a broader context.

The riverbed that we now call Yellowhouse Draw has revealed itself as an ancient interstate for animals and early humans crossing the region. Though it may seem incongruous to associate the dusty, windblown South Plains with water, it has always been present in the draws, attracting animals and humans. And while the 2.6-million-year-old animal fossils are vital to the Lubbock Lake Landmark's history, the 13,000 unbroken years of human activity are the foundation of its significance.

The Landmark's story of human occupation unfolds in five uninterrupted chapters, each preserved in distinct layers, or strata, of sediment and soil shaped by environmental and geological changes. When viewed in cross-section, the strata resemble a layer cake. Wind and water buried the bones and artifacts (human-made materials) within each layer, preserving the Landmark's extraordinary record. The Landmark's multiple distinct cultural layers within the strata stand in contrast to many other archaeological sites, which typically reveal only one or two cultural layers of human activity. Often, those layers are not preserved within well-defined geologic strata, making it difficult to determine the age of a specific artifact.

While other archaeological sites in North America have earlier evidence of human presence, none rival the Landmark's consistent layers of human history. "Any one of the remarkable records of plant and animal communities, geology, climate, environmental resources, or human activities would make Lubbock Lake significant, but we have it all," says Vance Holliday, a retired professor of geology at the University of Arizona and a longtime colleague of Johnson's. Holliday volunteered during the first excavation season led by Johnson at Lubbock Lake in 1973 and was among the first graduating class of Texas Tech's master's degree program in museum science. Now a noted expert in geology and Paleoindian archaeology, Holliday has collaborated with Johnson over the years and has maintained his interest in Lubbock Lake throughout his career.

Era	Period	Date
Cenozoic		**66 million years ago to present**
	Quaternary	2.6 million years ago to present
	Holocene	11,000 years go to present
	Pleistocene	2.6 million years ago to 11,000 years ago
	Neogene	23 to 2.6 million years ago
	Paleogene	66 to 23 million years ago
Mesozoic		**252 to 66 million years ago**
	Cretaceous	145 to 66 million years ago
	Jurassic	201 to 145 million years ago
	Triassic	252 to 201 million years ago
Paleozoic		**541 to 252 million years ago**
	Permian	298 to 252 million years ago
	Pennsylvanian	323 to 298 million years ago
	Mississippian	358 to 323 million years ago
	Devonian	419 to 359 million years ago
	Silurian	444 to 419 million years ago
	Ordovician	485 to 444 million years ago
	Cambrian	541 to 485 million years ago
Precambrian		**4.6 to 541 billion years ago**
	Proterozoic	2.5 billion to 541 million years ago
	Archean	4.0 to 2.5 billion years ago
	Hadean	4.6 to 4.0 billion years ago

Figure 2. Geologic timeline. (Based on a National Park Service graphic)

IN THE BEGINNING

The primary focus of the Landmark researchers is the Quaternary Period, which began about 2.6 million years ago. The Earth's continents had shifted into their current positions, and the Southern High Plains landscape had stabilized.[1]

The Landmark's story begins, however, much further back in time. What we know today as the Lubbock Lake Landmark can trace its origin to a major seismic event approximately 60 million years ago, which shaped the Rocky Mountains. As the mountains rose, erosion carried sediments as far east as Kansas and Nebraska, ultimately creating the Ogallala Formation.[2] The Llano Estacado also began taking shape on the Texas South Plains and the Eastern New Mexico Mescalero Plain.

The Ogallala Formation, formed between 23.5 and five million years ago, spans eight states from Wyoming and South Dakota to Texas. Within it lies the Ogallala Aquifer, one of the largest aquifers in the United States. It remains a primary water source for parts of the Great Plains, extending from South Dakota through Texas.[3]

The Blanco Formation, which is the Landmark's bedrock, developed between 2.58 and 1.6 million years ago. It lies in a vast ancient lake basin on the eastern edge of the Llano Estacado. The lake basin is cut into the Ogallala Formation and, at one time, stretched from present-day Lubbock to Crosbyton and Blanco Canyon.

Near the top of the Ogallala Formation, an ancient soil in the upper sediments hardened into a tough and thick calcrete layer, which we now know as the Caprock. The Caprock lends the Llano Estacado its flat appearance. When broken up, the calcrete produces caliche, a common component of road construction today.[4]

Understanding the area's stratigraphy allows researchers to date the bones and artifacts they uncover at the Landmark and other regional sites. Many of the sediments and artifacts found at regional excavations closely resemble those at the Landmark, establishing it as a touchstone or reference point for evaluating other sites.

However, it is the Landmark's five distinct stratigraphic layers that enable researchers to trace the connections that link our prehistory to our present, forming an almost complete cultural sequence that stretches back more than 13,000 years to the Clovis Period. The discovery of so many well-defined, intact sediment

Figure 3. Vance Holliday leans against an exposed wall that shows the different strata at the Lubbock Lake Landmark. (Courtesy of Lubbock Lake Landmark)

layers with minimal erosion is rare, making the Lubbock Lake Landmark one of the most significant archaeological sites in the Americas.

CHAPTER 2

LAYERS OF HISTORY

The initial suggestion of ancient human habitation at Lubbock Lake, then known as Long Lake, surfaced in 1936. By that time, the City of Lubbock's growing population required new water sources. The city had owned the area around the lake since 1903, when it bought the 93-acre lot from a homesteader for ninety-three dollars as a potential water source.

City officials knew that a freshwater stream had flowed through Yellowhouse Draw. Hoping the spring was only buried under silt, city leaders believed clearing it might restore the water flow and solve their supply issues. Using funds from the Works Progress Administration (WPA), they hired workers to dig the area by hand. That effort resulted in a swampy mess, compelling the city to use heavy equipment to dredge the site. Eventually, workers struck water, creating a U-shaped reservoir approximately 1,300 feet long and as much as 20 feet deep, which eventually filled with clear water. Ultimately, the city decided to use the new reservoir as a dedicated water source for the fire department.[1]

The dredging left behind large mounds of dirt, creating a perfect spot for curious local kids to explore. While searching through the dirt, two teenage boys discovered what they believed were arrowheads and bones. They brought their finds to William Curry Holden, the director of the West Texas Museum, which was founded in 1929 on the Texas Technological College campus. Holden, who joined the college the same year to teach history and anthropology, recognized the objects as Folsom-era points and ancient bison bones. Holden was familiar with earlier Folsom and Clovis discoveries.[2]

A Colorado Museum of Natural History (now the Denver Museum of Nature and Science) team began work in 1926 on a site discovered in 1908 near the small town of Folsom, New Mexico. They identified bones of several extinct ancient bison (*Bison antiquus*) dating from 12,800 to 12,300 years ago. In 1927, they made one of the most important discoveries in American archaeology: a stone spear point in situ (in its original place) among bison bones. The find marked the first confirmed evidence of human-made artifacts associated with Ice Age mammals in the Americas.[3] Finding a bone or artifact in situ is crucial to accurately dating an object.

Two years later, excavations near Clovis, in Eastern New Mexico, uncovered tools associated with a mammoth kill dating back 13,500 to 12,800 years ago. Scientists found bones and artifacts beneath Folsom levels, suggesting that humans lived in the region earlier than previously thought. For decades, scientists believed Clovis people were the first to inhabit the New World.[4]

Since the first accidental discovery of Folsom points at Lubbock Lake, researchers have known it holds layers of invaluable prehistory. They did not, however, realize the extent of the treasure buried within those layers.

FIRST EXPLORATIONS

Archaeological explorations have taken place across the Llano Estacado since the mid-1920s. The Llano Estacado stretches from

Figure 4. During Joe Ben Wheat's first excavation of Lubbock Lake in 1939, the high water table within the lake prevented him from reaching the bones visible at the lake's bottom. (Courtesy of Lubbock Lake Landmark)

Eastern New Mexico to east of Lubbock, where the Caprock ends, and includes both the Folsom and Clovis sites.

Work at Lubbock Lake commenced in 1939, when Holden recruited graduate student Joe Ben Wheat to excavate the site. The WPA financed Wheat's first excavation, which unearthed fossils and artifacts, underscoring the site's importance. Wheat later became the Curator of Anthropology at the University of Colorado Museum and a leading expert on Paleoindian studies.

Ironically, though the 1929 dredging damaged some fossils and artifacts, it revealed an archaeological gold mine of bones extending from the reservoir floor and sides. While Wheat could see the bones, the water in the reservoir prevented him from reaching them. Focusing on areas around the reservoir's rim, Wheat recovered artifacts and bones that scientists eventually associated with the Archaic Period, dating from 9,500 to 1,950 years ago. He was also able to identify at least five distinct stratigraphic levels, the first hint of the true significance of the site.[5]

Wheat's work set the stage for future excavations. Glen Evans, a geologist with the Texas Memorial Museum in Austin, was interested in Paleoindian sites and heard about the discovery at Lubbock Lake. Evans and his colleague E. H. Sellards, director of the Bureau of Economic Geology at the University of Texas in Austin, would talk with Holden in 1937 but did not begin any work at the site until the late 1940s.[6]

The decade after Wheat's excavations, the water table fell, eventually draining the reservoir. With the water gone, what Holden described as beautiful geological banding of the almost vertical banks emerged.[7] Those layers would eventually reveal an unbroken chain of human habitation dating from Paleoindian to Historic times. The reservoir would remain dry until 1979, when the water table rose, making digging inside the lake's banks impossible.

Evans and Grayson E. Meade, a geology professor at Texas Technological College, conducted the first fieldwork within the reservoir's banks in 1948 and 1951. Their efforts established the site as home to Clovis-era people, cementing the significance of the Lubbock Lake Landmark. Evans and Meade discovered a stone spearhead on the dredge deposits bearing the distinctive marks of the Clovis Period (13,500 to 12,800 years ago), as well as the bones of now-extinct Columbian mammoths (*Mammuthus columbi*) and ancient bison (*Bison antiquus*).[8] That discovery pushed back the established presence of humans at Lubbock Lake by about one thousand years. The duo's further excavations reinforced the existence of Clovis humans with the discovery of a tool and mammoth bones beneath the Folsom layer in gray sand similar to that of the Blackwater Draw Clovis site. These findings concluded that Clovis people had entered the area before the Folsom-era occupation.

Evans and Meade's work was groundbreaking on another front. They discovered Folsom artifacts and extinct bison bones in situ. The pair would send charred bones to Sellards, with whom they collaborated on one of the first uses of radiocarbon dating

Figure 5. Archaeologist Jane Holden Kelley (left), daughter of Curry Holden, talks with Eileen Johnson during an Archaeology Day at the Landmark. (Courtesy of Lubbock Lake Landmark; Kippra D. Hopper, photographer)

for Paleoindian sites.[9] The testing revealed that the bones were approximately 11,220 years old, which at the time was regarded as the earliest date recorded for Paleoindian artifacts. Though researchers initially believed the early testing indicated a Folsom bison kill, more recent research demonstrates that the bones date to the Plainview Period, about 11,500 years ago.[10]

In 1959 and 1960, Earl Green and Jane Holden Kelley, Curry Holden's daughter, working through the West Texas Museum, obtained a National Science Foundation grant to conduct excavations in the upper levels of what was then called the Lubbock Reservoir site. Working within the reservoir cut, they undertook further fieldwork and attempted to gather and catalog materials researchers had discovered before 1961. Green was then the director of the museum. Kelley eventually left the museum for positions at the University of Nebraska and would later become an assistant professor of anthropology at the University of Calgary, Canada.[11]

Excavations at Lubbock Lake were sporadic until Eileen Johnson arrived in 1973 to oversee the work there. She began yearly

Figure 6. Eileen Johnson's first excavation in 1973 appears in the background. From left are Craig Black, then director of the Museum of Texas Tech University; Jack Hughes, from West Texas State University; Curry Holden; and Joe Ben Wheat. (Photo courtesy of Lubbock Lake Landmark)

excavations that continue today. She also expanded the Landmark's work to a variety of regional sites. The 1973 summer field season, conducted through the Museum of Texas Tech University (the new name adopted in 1969 for the West Texas Museum), explored several locations around the reservoir to determine which areas might prove most productive for future work.[12]

Work at two places around the reservoir's edge proved highly successful, yielding numerous bones and hand-fashioned stone tools. The excavation crew found signs of human habitation in

Period	Date
Historic	380 years ago to present
Protohistoric	530 to 380 years ago
Ceramic	1,950 to 530 years ago
Archaic	9,500 to 1,950 years ago
Firstview	9,500 years ago
Plainview	11,500 years ago
Folsom	12,800 to 12,300 years ago
Clovis	13,500 to 12,800 years ago

Figure 7. Landmark cultural periods. (Based on a graphic from Lubbock Lake Landmark)

the sediment layer dating to the Archaic Period (9,500 to 1,950 years ago); ancient bison bones showing butchering marks in the Folsom level; and mammoth and ancient horse bones embedded in the Clovis-level sands. These clues hinted at the extent of ancient riches buried at Lubbock Lake and guided Johnson's team to focus on those two locations for the 1974 season. The discoveries also bolstered their conviction to protect and preserve the site.[13]

CULTURAL HISTORY

The Lubbock Lake Landmark anchors one of the most significant archaeological sites in the New World, holding a unique place in the story of human history in the Americas. A living textbook, the site tells the unbroken story of how humans, animals, and the landscape have evolved over time, in five distinct chapters, or layers.

The earliest human evidence at Lubbock Lake Landmark dates back to the Clovis Period, the earliest recognized cultural period on the Southern High Plains. This period is named after Clovis, New Mexico, where archaeologists first discovered artifacts from this era.

From the Clovis Period onward, the site's stratigraphy holds proof of a continuous sequence of cultural periods that define human presence in North America. Scientists refer to the earliest humans as Paleoindians (ancient Indians). Distinctive markings on stone spear points often identify Clovis archaeological sites. Clovis stone spear points have been described as made from jasper, chert, obsidian, and other fine, brittle stone. They have sharp edges and grooves, called flutes, extending from the base toward the tip. Typically, the points are roughly four inches long and a third of an inch thick.[14]

One Clovis point and an intact Clovis-era processing station were found in Stratum 1, the oldest layer identified at the Landmark. Stratum 1 is divided into three substrata; the oldest rests on the Blanco Formation, the bedrock at the Landmark. The processing station also held the bones of extinct bison, Columbian mammoth, horse, camel, and bear. Radiocarbon dating of a preserved hackberry log in Substratum 1B yielded a calendar date of approximately 12,990 years ago.[15] From the gravel and sands in Stratum 1, researchers know that a river flowed through Yellowhouse Draw for thousands of years. The stream became less active, and plant and animal remains indicate that winters were mostly mild and summers mostly cool, with the landscape marked by meandering streams and marshy meadows.

Stratum 2 has two substrata and contains evidence of the Folsom Period, dating from approximately 12,800 to 12,300 years ago. Lubbock Lake was initially best known as a Folsom site, but the discovery of Clovis-era material and later Paleoindian artifacts shifted that assumption. Burned bison bones recovered by Evans and Meade in the 1940s and radiocarbon dating in the 1950s provided the first Folsom-era date. Johnson notes that Landmark scientists now know that the Folsom occupation at the Landmark is confined to Substratum 2A, the lowest level of Stratum 2. In retracing Evans and Meade's work, she says, the burned bone from a bison kill site from upper Substratum 2B is in what is known as the Plainview Age.

Folsom projectile points are recognized by distinctive fluting along the length of the point, as represented by the two Folsom points recovered from the dredging of the springs in 1936. Most artifacts from this period are found with remains of ancient bison, the chief food source for early humans. Evidence suggests that people used the bones as tools, which they employed on-site to skin animals and strip the meat from the bones for food.[16] Stratum 2 indicates winters were cooler than before and summers were becoming warmer. The landscape was a prairie habitat with ponds interconnected by a stream.

The upper layer of Stratum 2 preserves evidence of the Plainview and Firstview cultural periods. The Plainview Period spans approximately 11,500 to 9,500 years ago, followed by Firstview, which dates to around 9,500 years ago at the Landmark. Plainview people used butchering methods similar to those of the Folsom Period inhabitants.

During the Firstview Period, butchering methods began to change. Instead of stripping an entire carcass, hunters began selecting specific cuts of meat. As in earlier periods, bones were still made into tools and left with the remaining butchering debris. Projectile points from both periods have parallel sides that taper to a point, differentiating them from earlier types. The climate mainly remained unchanged, with marshy areas transitioning into open prairies.

The Paleoindian Period gives way to the Archaic Period, which varies in duration across North America, from approximately 10,000 to 1,950 years ago, and is divided into three phases: Early, Middle, and Late. All three are represented at the Landmark, although none is well-defined at the site or elsewhere across the Llano Estacado. The climate became warmer and drier. A nearly one-thousand-year-long drought followed a period of freshwater marshes.

Archaeologists have recovered a few Archaic-era projectile points on the surface and in situ at Lubbock Lake in Strata 3 and 4.

However, the scarcity of artifacts and bones makes it difficult to determine how people lived during this period. Scientists agree that the Archaic is a transitional period. Giant Ice Age mammals had long since gone extinct, whereas many smaller mammals, including bison and wolves, survived and evolved into the creatures we know today. People continued to be hunter-gatherers, with bison remaining the primary food source.

The Ceramic Period began approximately 2,000 years ago and ended around 500 years ago. Straddling Strata 4 and 5, this period shows the beginning of a cyclical drought pattern. Spring-fed marsh-and-stream complexes increased in size. Excavations at the Landmark have yielded small projectile points that signal the first use of bow and arrow technology, animal-kill sites, and camping areas. For the first time, archaeologists found ceramic materials at the Landmark. During his 1939 excavation, Joe Ben Wheat discovered sherds of Chupadero black-on-white pottery, made by the Jornada Mogollon, an ancestral pueblo people from north-central New Mexico in the upper reaches of Stratum 5, dating from approximately 950 to 1,300 years ago.[17]

During the Protohistoric Period (500 to 300 years ago), the lowest level of Stratum 5 shows evidence of repeated use by the early Apaches, who migrated from Canada between 600 and 650 years ago. Excavation crews have found both camping and butchering sites. A key indicator of this period is the Garza point, a small triangular point notched at the bottom. Stratum 5, the latest of the Landmark's layers, holds pottery sherds identified as Apache. This period also saw the arrival of Europeans on the Llano Estacado. There are indications of widespread marshlands, and animal remains suggest a grasslands habitat.

The Historic Period (300 years ago to the present) reflects the influence of European peoples. Crews have found modern horse bones, glass, and metal in upper Stratum 5. The ancient horse, which developed in the Americas and then migrated to

Europe and Asia, became extinct in the Americas around 13,000 years ago. The Spaniards reintroduced horses in the late 1500s. Early Apaches initially used horses as a food source but later adapted them for transportation. The Comanches, who displaced the Apaches in the early 1700s, became renowned for their expert horsemanship, utilizing horses for both hunting and transportation.

By the mid-1800s, westward expansion was well underway, bringing unprecedented change to the Llano Estacado and the South Plains. The completion of the transcontinental railroad in 1869 facilitated the shipping of bison hides east, which consumers valued making warm coats and blankets. Manufacturers also utilized the hides to produce stronger belts for machinery. Consumer demand, coupled with the US government's support of hunting of bison as a method of removing Native American tribes from the Plains, opened up the territory to settlers and ranchers.[18]

SINGER STORE

Among the early settlers on the Llano Estacado were George and Rachel Singer. Singer built a store in 1883. Originally located near the bank of Long Lake, it was the first retail establishment in what would become Lubbock. What remains of the structure has yielded a wealth of cultural material and a mystery. Singer and his family had joined other Quaker families in moving west in 1880. Singer arrived at the end of the buffalo-hunter and *pastores* (sheepherders from what is now New Mexico) occupations of the area, and at the beginning of the early cattle ranches. Wisely, Singer chose a spot near the water that Long Lake provided. The lake served as a stopping point for cattle drives traveling north. It was also located on the axis of two military trails. Soldiers moving east to west between Fort Griffin and Fort Sumner and north to south from Fort Elliott to Fort Stockton traveled past his store.

Figure 8. Excavation around the Singer store location at the Landmark. (Courtesy of Lubbock Lake Landmark)

Catering primarily to cowboys from nearby ranches and traders, the Singer store stocked foodstuffs, including sugar, flour, canned meat, peaches, and beans. The establishment also carried boots, saddles, tobacco, blankets, and barbed wire.

A fire destroyed the original store in 1886. The blaze started during a fight between Singer and a Mexican stranger. Johnson could not find anything about the stranger until 2021 or 2022; she doesn't remember the exact date. A colleague sent her a clipping from the April 26, 1886, *Galveston Daily News*, the Galveston, Texas, newspaper reporting the fight and the fire. The article notes that a group of freighters picked up a Mexican man along their route and made camp about two miles from the Singer store. The man stole the freighters' guns and ordered them to leave, firing at them as they fled. Two teamsters went to the store the following day and told Singer what had happened. Later, the Mexican man showed up and recognized the two. He shot at the men and Singer. Singer apparently ran to his nearby house, got his gun, and returned fire. It was then that smoke was seen coming from the store. When the fire was out, the Mexican was found dead in the debris near the front door of the store. The newspaper article notes that while his limbs were charred, his face was only slightly burned. The article goes on to suggest that "he suicided rather than be suffocated or take the chances of being killed or captured should he attempt to leave the building." It is also not known whether the man set the fire or if Singer's bullet hit something flammable.[19]

Johnson and her teams have spent years excavating the area and have found an array of items, including a woman's hairpin, horseshoes, nails, and other objects. What frustrates her is that she has never found the Singer home. There are historical references to the house but conflicting information about its location in relation to the store. "It had to have been a sizeable structure," Johnson says, noting the family had several children. "It's strange we have found no indication of the house." Excavation near the store site continues, and she hopes that future digs will reveal the house.

HOW AND WHEN EARLY HUMANS ARRIVED

Curry Holden predicted in 1939 that the Landmark would play a significant role in identifying the extent of human habitation on the Llano Estacado. He was correct. Johnson notes, however, that the questions of *when* and *how* humans arrived in North America remain controversial.

Science evolves as researchers uncover new information. For example, Johnson notes that when she first entered the profession in the early 1970s, scientists generally agreed on the timing of human arrival in the New World. "We believed that they arrived exactly 12,800 years ago, and they hunted mammoths," she says. "Those assumptions don't hold up today. Yes, they did hunt mammoths, but they also hunted other big animals."

For decades, archaeologists believed that humans first arrived in the New World around 13,000 years ago. They based this assumption on the Clovis points discovered near Clovis, New Mexico. Other excavations, such as the 1997 Monte Verde, Chile, excavation, indicate that humans inhabited the Americas about 14,500 years ago, based on radiocarbon dating.[20] Other sites in Oregon, Idaho, and Texas pushed the human habitation date back to 16,000 years ago.[21] Johnson notes that today, most researchers are comfortable supposing that humans arrived about 18,000 years ago.

Now, a more recent discovery at White Sands National Park in New Mexico could push back the date of the first human occupation to 21,000 to 23,000 years ago, calling into question how humans arrived in North America, south of the glaciers that covered much of present-day Canada and the northernmost points of America. The initial dates announced in 2021 relied on radiocarbon dating of seeds. In 2023, researchers confirmed the dates through radiocarbon dating of pollen.[22]

Vance Holliday has conducted Paleoindian archaeological work in the lakes at White Sands since 2012, investigating Paleoindian archaeology in the context of some ancient lakes. "I am not a footprint expert," he emphasizes. "But due to my earlier involvement in the area, I was invited to get involved in the research into those tracks." He notes that while there may be thousands of footprints made by large animals in the park, there are not many made by humans. "In the excavated and dated site, there are about sixty individual footprints and eight trackways."

Holliday has remained involved and in June 2025, he and colleagues published a paper in *Science Advances*, the American Association for the Advancement of Science's (AAAS) open-access multidisciplinary journal. The paper presents new research and radiocarbon dates from the site, conducted by researchers and laboratories independent of the work presented in 2021 and 2023. The new dates were determined on a third source of carbon, wetland

Figure 9. The discovery of human footprints at White Sands National Park in New Mexico could push back the acknowledged date of the first human occupation to 21,000 to 23,000 years ago. (Courtesy of Vance Holliday)

deposits, which confirm the dates on seeds and pollen in the original research. The 2021 paper concludes that the White Sands dates indicate that humans arrived in the Americas before the Last

Glacial Maximum, a time when ice covered North America from the Northern Atlantic to the Northern Pacific. The ice would have cut off travel to North America from Asia, the route most generally accepted as the one that ancient humans took to America. The arrival of the first people in the Americas has long been a topic of controversy. While the paper supports the original White Sands dates, it will most likely not end the debate.[23]

Even though she and Holliday are longtime friends and colleagues, Johnson says she remains cautious about the original dates. "Vance's paper, nevertheless, sets a strong foundation for White Sands eventual acceptance," she notes. "However, it will take time and the discovery of more than one location dating to this time to build scientific consensus." As for Holliday, he has no problem with Johnson's guarded approach, noting that "science would be boring if everyone agreed on everything."

While White Sands is close to Lubbock, about 300 miles west by today's highway route, Johnson says that even if the White Sands dates are accurate, she does not believe pre-Clovis people would have visited Lubbock Lake. The draws on the Southern High Plains had not yet been cut down. "That began around 21,500 calendar years ago, at the end of the Last Glacial Maximum," she says. "The Salinas [a salt marsh or salt flat] would have existed, and maybe some playas [dry lake beds at the bottom of a desert basin, sometimes temporarily covered with water]. If people were in the region as early as the White Sands dates, then that is where the occupations would have been. It's an interesting, complex situation."

Just as the question of when humans arrived in the New World remains controversial, so too does the question of how. The prevailing theory centers on the landmass known as Beringia, which once spanned the Bering Strait, extending from Siberia to Alaska. During the last ice age, receding sea levels exposed this land bridge, allowing animals and humans to migrate into the New World. A

2022 study published in the *Proceedings of the National Academy of Sciences* estimates that Beringia was accessible between approximately 46,000 and 35,700 years ago. The research also indicates that humans migrated to the Americas as soon as the land bridge formed. Some scientists believe humans walked southward through glacier openings as the latter began retreating 14,000 to 15,000 years ago.[24]

Though there is no solid archaeological evidence, one theory suggests that discoveries of human habitation earlier than 15,000 years ago indicate that humans reached the Americas by means other than walking across the Bering Strait. The Trans-Pacific Migration Theory posits that humans arrived by boats along the western coast. An article on the History.com website, which is sponsored by the A&E Networks, the parent company of the History Channel, notes that the western coastline of North America was ice-free thousands of years earlier than the continent's interior. Pre-Clovis humans might have used boats to journey to North and South America.[25] If humans did walk from present-day Alaska to the Llano Estacado, it is not known how long it took them to arrive. Johnson estimates the journey took several hundred, but most likely less than one thousand, years.

Another unanswered question is how early people navigated their new and different surroundings. While early humans were accustomed to living with large animals in Siberia and Asia, the vegetation and environment they encountered on the Southern High Plains would have differed significantly, Johnson explains. They would have crossed prairies, and the area's temperature and humidity would have been different. Those first people would have had to learn to adapt quickly, and Johnson wonders if and how they might have shared information with those who came after them. "If Clovis people were not the first into our area, did the earlier people set up territories? If so, how permeable were those boundaries?" she asks. "We're looking at how they gained knowledge that they needed to live, and if they passed it on to others coming through the area."

However and whenever Paleoindians first arrived in the Americas, Johnson believes they resembled present-day Native Americans. She explains that her reasoning is based on reconstructions of skeletons from early burial sites. "Ancient DNA indicates they most likely had brown eyes and darker skin," she says. "I've read studies that report the blue eye mutation developed six to eight thousand years ago."

EARLY ANIMALS

Some ancient animals evolved in North America; others, such as early humans, likely crossed the Bering Strait land bridge between Siberia and Alaska, as well as the Isthmus of Panama in the south, before migrating to North America. While the Lubbock Lake Landmark's significance lies in its continuous human record, it has also yielded an array of animal and plant fossils that contribute to our understanding of how the landscape changed and how humans and animals coexisted.

The more exotic animals, such as Columbian mammoths (*Mammuthus columbi*), native camels (*Camelops*), horses (*Equus*), and giant short-faced bears (*Arctodus simus*), are perhaps the most interesting to the public. The Landmark's collection also contains the remains of thousands of plants and creatures, including fish, mice, snakes, and birds. "When you look at the total collection from the Landmark, you get a much richer understanding of the ecology through time," Johnson says.

Mammal records on the Southern High Plains stretch back millions of years. At the Lubbock Lake Landmark, excavations have recovered a small but significant collection of plant and animal fossils dating back about two million years from the Blanco Formation. By around 13,000 years ago, the earliest confirmed date at the Landmark, vast grasslands surrounded the area.

The woolly mammoth (*Mammuthus primigenius*) and Columbian mammoth are perhaps the most recognizable Ice Age creatures. Fossils of woolly mammoths, which were more suited to the icy cold of the northern part of North America, never appeared on the Southern Plains. Columbian mammoths, on the other hand, were better suited to the warmer southwestern environment. As they were grazers, the grasslands likely offered a perfect habitat.

The creatures were enormous, with adult males standing thirteen feet at the shoulder and weighing approximately 22,000 pounds. In comparison, today's African elephant ranges from eight to thirteen feet at the shoulder and weighs 5,000 to 14,000 pounds. The contemporary Asian elephant stands six to twelve feet at the shoulder and weighs 6,000 to 12,000 pounds. Yet, Johnson explains, the Columbian mammoth is more closely related to Asian than African elephants. "That is a bit of a DNA shocker," she acknowledges.

Columbian mammoth shared the grasslands with camels and horses, both native to the Americas, as well as the ancient bison. Johnson explains that tracing the horse family tree is a complicated process. Equidae, the taxonomic family of modern horses, zebras, and other related animals, evolved about fifty million years ago in the North American grasslands. The genus *Equus*, to which modern horses belong, evolved four to five million years ago. "The *Equus* horses went extinct in the New World at the end of the Pleistocene 12,800 years ago, as did most large mammals," Johnson explains. "The *Equus* horses lived on in the Old World. Those horses gave rise to the modern domesticated horse and were the horses the Spanish brought back to the New World." She notes that the horse family tree has more than a dozen branches, but DNA shows only two to four types of late Pleistocene horses that she believes inhabited the North American grasslands.

The camels that made the Llano Estacado their home were much the same size as those of today and resembled their modern cousins. They originated in North America and eventually migrated to Asia. Scientists cannot determine whether the camels had humps because the fatty tissue that would have formed that feature would not have been preserved as bone was.

Landmark excavations have revealed that ancient bison evolved into modern bison sometime after 8,000 years ago and before 6,500 years ago. Modern bison survived a near-extinction event that happened from the 1870s to the early 1880s, due to commercial hunting and US military policies. The ancient bison were about 25 percent larger than today's bison (*Bison bison*). Standing about seven-and-a-half-feet tall and weighing around 3,500 pounds, the ancient bison had longer horns than today's bison.

Ancient and modern bison were primary food sources for people around the Landmark. Very little grass is preserved in the archaeological record. However, the presence of bison, an herbivore, at the Landmark strongly indicates that the area was a grassland. Both ancient and modern bison belong to the bovine group, which also includes cattle.

Giant short-faced bears also roamed the area. They were genuinely fearsome-looking creatures, measuring five feet at the shoulder and up to twelve feet when standing on their hind legs. Scientists believe they ran faster than any other bears. Primarily a carnivore, the giant short-faced bear wasn't a great hunter. Johnson says that research shows the bear was most likely a scavenger, subsisting on the kills of other predators.

Life-size bronze replicas of the bear, bison, mammoth, and giant pampatheres (*Holmesina septentrionalis*), a relative of the modern-day armadillo, surround the Nash Interpretive Center at the Landmark. These creatures, along with most large mammals, became extinct approximately 12,800 years ago. As with previous extinction events, different mammals subsequently emerged. Many were smaller than the Ice Age creatures. A lucky few, such as prairie dogs, wolves, and modern bison, remain with us today.

MORE TO FIND

Continuous efforts to preserve the Landmark have encountered varying degrees of success. Urban development, road construction, and misguided public policies have caused severe damage over the decades. Though the Landmark spans more than three hundred acres, Johnson knows there is much more to explore in the greater region if her crews could have access. Yet, much of the land surrounding the site belongs to the City of Lubbock or private landowners. Although relationships with the city have not always been smooth, officials have invited Johnson to conduct state-mandated archaeological reviews of municipal projects. One project involved a new pipeline that identified potentially rich archaeological sites. While her study led the city to revise its pipeline plans, she was unable to access the area for further exploration.

When the Texas Department of Transportation (TxDOT) conducted a similar archaeological survey before constructing Loop 88 to the west and south of Lubbock, officials requested Johnson's assistance. The department announced in May 2025 that the excavation had unearthed Ice Age fossils, specifically a tooth of an ancient giant ground sloth and bones from two other animals.[26]

Johnson says finding remains of the ancient creature is incredibly rare. Johnson's team at the Landmark is preserving and studying the fossils. While excited by the find, she warns that it will take at least a year, and possibly longer, to identify the fossils from the excavation. The remains are fragile, and months will be needed to stabilize the bones so that she can begin work to determine the species.

Several types of ground sloths existed during the Ice Age, all of which were huge. Johnson notes that only three types of ground sloths lived in Texas. *Megalonyx jeffersonii*, named after President Thomas Jefferson, was eight to ten feet in length and weighed 2,200 to 2,400 pounds. *Paramylodn harlani* was primarily a grazer that inhabited grasslands. It was approximately ten feet in length and weighed 2,400 to 3,000 pounds. The smallest of the ancient sloths was the *Northropes shastensis*, which measured 9 feet from snout to tail and weighed only five hundred pounds.

The TxDOT excavation is an example of the yield of treasures waiting to be discovered on the Llano Estacado. Johnson knows that the cultural record extends north and downstream from the Landmark, but she says development destroyed the record. "We have a sizeable portion preserved. The Landmark is one of the largest hunter-gatherer locations in the entire New World," Johnson explains. "Human habitation here is not just one time; it has occurred repeatedly over thousands of years. The Landmark is an amazing complex, and we're fortunate to have the chunk we do."

CHAPTER 3

THE POLITICS OF PRESERVATION

Lubbock Lake is a treasure chest of ancient artifacts and bones. In a sense, the Landmark and its sister regional excavation sites are time capsules buried by Mother Nature, offering us glimpses into the lives of the animals and humans who once roamed these plains and the significance of their stories. But preserving those treasures has not been easy.

The 1972 hiring of Craig Black, PhD, as director of the Museum of Texas Tech University set in motion more than half a century of continual research at Lubbock Lake Landmark. Texas Tech hired Black away from the University of Kansas, where he was Curator of Vertebrate Paleontology in the university's Museum of Natural History. He was a well-known, highly regarded vertebrate paleontologist.

Black became Eileen Johnson's mentor. She met him in 1970 on her first day on the KU campus when she walked into the museum and asked to speak to the paleontology curator. "I sat down and asked him for a job," she laughs. "It happened that there was a student assistant position. It turns out it was Craig's first day on the job."

Johnson says she was floored when Black told her he was leaving Kansas. She did not want him to go. Her perspective changed, however, when he offered her an opportunity that would become her life's passion.

Black informed Johnson that a site affiliated with the Museum of Texas Tech needed "to be looked at." When he said it was Lubbock Lake and asked if she thought she could do anything with it, Johnson recalls that she almost fell off her chair. Although the museum participated in work at the site, it was owned by the City of Lubbock. Archaeologists around the country already recognized the importance of Lubbock Lake. "I had heard of the site and couldn't believe what I was being handed. Craig wanted me to 'do something with it,'" Johnson says. "And here I still am trying to do something with it despite more than fifty years of unearthing the Landmark's history."

She enrolled in the zoology doctoral program at Texas Tech, thinking she would spend about three years researching at Lubbock Lake, writing her dissertation, and then moving on. She earned her doctorate but never left Lubbock. "After that first field season, I knew I wouldn't be finished in three years; I'm not finished now," Johnson says.

Though the research remains her passion, she moved into teaching in 1981 when Texas Tech hired her as an assistant professor in Museum Science, a master's degree. She would later become chairperson of the program, a position she still holds as of 2026. While she has always overseen the excavations at Lubbock Lake, she served as a graduate student or research assistant until 1981, when she officially became Lubbock Lake Landmark director. Johnson would grow the research at the locations and oversee the program's expansion to regional excavation sites. Perhaps her most important job: She would also become its protector.

HOLDEN THE VISIONARY

When William Curry Holden saw those first stone spear points in 1936, he recognized the archaeological importance of Lubbock Lake and knew the site must be preserved. He passionately believed that educating the community about its importance was vital, and he never missed an opportunity to take community leaders and anyone else interested on a tour of what he called the old watering hole or the old dig. He brought college students studying history and anthropology on field trips to Lubbock Lake. He encouraged students who explored the site independently to report any findings to him.[1]

Research continued sporadically at the reservoir. However, when a permanent water source failed to materialize after the dredging, the city had no plans for the land around the lake, leading to several archaeologically destructive events in Yellowhouse Draw. In the 1940s, crews dug four caliche pits along the draw near the reservoir, destroying bones and artifacts. Construction of several pipelines across the draw came close to the reservoir. In the 1960s, a new sewage treatment plant destroyed Holocene Period artifacts dating back at least one thousand years. Though the plant is no longer in service, the structure remains.[2]

Holden continued his advocacy for efforts to protect the area around the reservoir. In 1955, he made his first formal preservation effort when he requested that the Lubbock City Council officially recognize and preserve the site. Later that year, Holden again came before the council asking that the city lease the six acres immediately surrounding the reservoir to the West Texas Museum Association (WTMA) for ninety-nine years at $1 per year. At the time, those acres were what Holden thought most archaeologically crucial. Although the council approved a lease on five-and-a-half acres, the council did not sign the lease until 1958, designating the area as a historical, archaeological, and paleontological site.[3]

In 1966, WTMA erected a chain-link fence around the fifteen acres of excavations that archaeologists had established as the archaeological

Figure 10. The West Texas Museum Association erected a sign in 1966, hoping to draw interest to what was then called the Lubbock Lake Archaeological Development. (Courtesy of Lubbock Lake Landmark)

site's extent. WTMA installed large signs on the fence hailing what was then called the Lubbock Lake Archaeological Development and trumpeting the discovery of evidence of "Ancient Elephant Hunters" dating back twelve thousand years; the signs reflected Holden's desire to educate the public about what the site held.[4]

AN INVALUABLE PARTNER

Also in the 1960s, the Landmark gained a significant ally in Robert (Bob) Nash. He would become a prominent civic leader, ultimately serving as a city council member. He was an active member of the Westerners, a local history club; a popular Lubbock radio and television personality; and, eventually, the general manager of a local television station. Most important for the Lubbock Lake site, Nash was a passionate amateur historian.[5]

Lubbock County Judge Curtis Parrish began working at KAMC-TV in 1979 in the sports department before becoming chief photographer in the news department. Nash, meanwhile, was hired at KAMC in about 1980 as a television show host and became general manager in 1983.

Parrish remembers hours spent with Nash in the station's breakroom, listening to him talk about the area's history. He and Nash would arrive at the station at about the same time each morning. "We'd drink coffee, do the crossword puzzles, read the newspapers, and I would listen to Bob tell his stories," Parrish remembers. "It seemed that you couldn't talk to Bob for longer than two or three minutes before he would start talking about history. He had an immense love for the Llano Estacado."

Parrish recalls that one day in the late 1980s or early 1990s, Nash invited him on a trip to the edge of the Caprock near Crosbyton, about forty miles east of Lubbock. Nash led him to a spot overlooking the area off the Caprock that Parrish describes as triangular, worn smooth, with bits of dark rock chips in the center. "Bob sat down cross-legged and told me this was a spot where some Native American hundreds of years ago probably sat making arrowheads while he watched the land below him," Parrish says.

In the mid-1960s, the Lubbock Chamber of Commerce established its Goals for the '70s Committee, chaired by Nash. The committee's final report prioritized the preservation and development of Lubbock Lake to increase tourists' spending on the site and improve the city's economy.[6]

A COOPERATIVE ENDEAVOR

In 1968, the City of Lubbock and the Lubbock Chamber of Commerce set in motion a partnership with the West Texas Museum to develop the Lubbock Lake Site Park. While the museum, with Earl Green, PhD, as its director and one of the reservoir's early researchers, focused on the area immediately surrounding the site, the city was looking at what to do with the remainder of the land.

Figure 11. Field crew of researchers Earl Green and Jane Holden Kelley (L–R): Demetro León, Lazara Oliva, Jesús Mininez, and Manuel Vargas. The vehicle, which the crew used for fieldwork, was still in service in 1973 and served as Eileen Johnson's first field truck. Photo taken during 1961 excavations. (Courtesy of Lubbock Lake Landmark)

Though the city had always acknowledged the importance of archaeological research, the development of the remaining area almost always focused on attracting tourists to boost the city's economy. The city approved hiring a landscape architectural firm to plan a park surrounding the Lubbock Lake Site in May 1967. By October, the planned park received bad news when the city's Parks and Recreation Department director told the city council that funds to develop the park might not be available. The October 13, 1967, issue of the *Lubbock Avalanche-Journal* reported that when asked how much money the city could allocate in an upcoming bond election for the proposed park, A. C. Hamilton replied, "It's not a good picture as far as the next five years are concerned."[7]

Despite the funding uncertainty, master planning for the park continued. The city had hired Schrickel, Rollins and Associates, landscape architects and civil engineers from Arlington, Texas, to design the park. In April 1968, Gene Schrickel presented a plan to the Lubbock City Council for what would be called Comanche Canyon. The planners viewed the archaeological site as an important part of the park. On April 5, 1968, the *Avalanche-Journal* reported that the planners acknowledged the significance of the archaeological finds at the reservoir and included an interpretive center for exhibits, support buildings for researchers, observation stations for visitors, and a botanical garden as part of the concept. The planners also suggested that the Singer store, the first retail business in Lubbock, be reconstructed on its original site near the bank of the reservoir. The newspaper article details several features for the remaining acres, including a frontier-style fort, a Native American village, a Comanchero Camp, a buffalo hunters' camp, a children's dinosaur-themed play area, and the restoration of an early Lubbock train depot with a small train for tours of the facility and a children's zoo. As he outlined the grand plans for the park, Schrickel pointed out that the proposal did encounter challenges, such as drainage issues and ease of access to the area.[8]

Though the construction of Comanche Canyon Park never materialized, Johnson points to the significance of including an interpretive center and research buildings in the master plan. It shows that Holden's continued advocacy kept Lubbock Lake and a new interpretive center on the city's and community leaders' radar.

GROWING SUPPORT

The early 1970s proved pivotal for Lubbock Lake. Bob Nash became the site's leading champion, and Eileen Johnson would arrive on campus in 1972. Through the West Texas Museum, she launched what was then called the Lubbock Lake Project, which she now remembers as part of her thinking that she would

Figure 12. Eileen Johnson and Vance Holliday with the sign announcing the Lubbock Lake Site Project and funders. (Courtesy of Lubbock Lake Landmark)

spend three years at Lubbock Lake researching and writing her dissertation.

Continuing with the Chamber of Commerce's goal to preserve and develop Lubbock Lake, city leaders created the Lubbock Lake Site Development Corporation (LLSDC) in 1971, with Nash leading the organization. In 1973, the LLSDC became a nonprofit organization focused on preservation, interpretation, and public education.

LLSDC's first steps included acquiring a new 29-acre lease on the archaeological site, replacing the five-and-a-half-acre lease that the West Texas Museum Association held. In a chapter in the 2016 book *Interpreting First Peoples Bison Kills at Heritage Parks*, Johnson recounts that the LLSDC, represented by Nash, talked to incoming museum director Black and received his agreement in principle to continue research and the preservation and education efforts that Holden had initiated.[9]

In 1971, LLSDC helped secure a National Register of Historic Places designation, which Johnson says called significant attention

Figure 13. Tours of the Landmark were conducted intermittently until 1973, when Johnson began yearly tours that continue today. (Courtesy of Lubbock Lake Landmark)

to Lubbock Lake's archaeological importance and the necessity of preserving the site. In 1973, Johnson directed the first of more than half a century of summer field seasons at Lubbock Lake. In addition, she began the first regularly scheduled public tours. Building on those Holden had given over the years, Johnson led the tours herself, taking the public past working excavation areas. The following year, field crew members took over as tour guides, a system that would last until the mid-1980s, when community volunteers took over. Johnson believed that giving visitors an inside look at what the crews were finding each day lent a sense of immediacy and intimacy. She continues to think that exposing visitors to actual fieldwork affords "not only a better understanding of the archaeological process but also an appreciation for historic preservation through a personal experience."[10]

Johnson's almost immediate recognition that the research at Lubbock Lake would keep her at Texas Tech for the long term, coupled with Nash's unwavering desire to protect and expand the research efforts, created a strong partnership that would benefit the site for decades.

By 1977, Johnson knew that what needed protecting extended well beyond the site's allocated acres. In that 2016 book chapter, she wrote that the extent of the site covered at least three hundred acres, all the land that the City of Lubbock owned for what was now called Yellowhouse Canyon Park. Though the city was reluctant to dedicate three hundred acres to archaeological exploration, Nash moved forward with acquiring national and state designations to help with preservation efforts.[11]

PROTECTION AND MORE DESTRUCTION

In 1977, the federal government declared the entire city-owned three hundred acres a National Historic Landmark. Four years later, Texas designated the site a State Archeological Landmark. With those protections, the Lubbock Lake Project morphed into the Lubbock Lake National Historic and State Archeological Landmark, colloquially known as the Lubbock Lake Landmark.

Despite the new state and federal protections, the city allowed the installation of signal towers and pipelines. It designated the northern portion of the now Lubbock Lake Landmark as a recreational off-road vehicle area (where activity continued until 1987). It built athletic fields and an access road. Each project caused significant damage to potentially substantial archaeological finds at the Landmark. Each of these actions made it difficult, if not impossible, to know precisely which strata an artifact had been deposited in or to determine the associations among the artifacts.[12]

Johnson's passion for the Landmark and the city's apparent lack of understanding regarding the area's cultural significance were evident during the construction of an access road. In the early 1980s, a city crew began cutting a road to the athletic complex. The tale of

Johnson's jumping in front of a bulldozer to halt the construction has reached almost legendary proportions in its retelling over the years. Unaware that the street was being built, Johnson admits she did run out and got in front of the bulldozer. "I just wanted to stop the work," she explains. "The poor worker didn't know what was happening; he was just doing what he was told. But the city knew they were doing the wrong thing."

Johnson called Jim Goss, then the museum director, to see if he had authorized the work. "I had not!" he said in an oral history interview for Texas Tech's Southwest Collection/Special Collections Library. "There had been no discussion with the university or the museum. We have 12,000 years of history there, and they just put a road across the site, violating state and federal laws."

With the designations of national and state landmarks in place, any work on the land must abide by state and federal antiquities laws. Goss met with city officials, who, he says in the oral history, told him it was city land and that they could do with it what they wanted. Goss says he informed the university president, but the university didn't want to "make waves with the city administration."[13]

Federal landmark designations come from the U.S. Department of the Interior through the National Park Service. Goss's next call was to Washington, DC, resulting in a short work stoppage to allow Johnson's team to save any antiquities they could along the road's route. Ultimately, the street was finished and is now known as Landmark Drive. Johnson laments, however, that the construction destroyed so much history.[14]

FUNDING BATTLES

As Nash worked on preservation efforts, Johnson fought a continuous battle to fund research and educational activities. Almost all funding came from grants and donations. Nash and LLSDC

Figure 14. In 1973, City and County of Lubbock officials often pitched in to work at the Landmark, including County Commissioner Arch Lamb (center). With Lamb are Craig Black, museum director (left) and Bob Nash, civic leader and major Landmark supporter. (Courtesy of Lubbock Lake Landmark)

were instrumental in securing some grants, convincing the city to provide partial funding for the first field season in 1973, and donating and moving a wooden building to the field camp. The building was later remodeled into a camp kitchen, a function it continues to serve. The city donated and moved a second building to serve as a laboratory during the 1974 field season. That building is now an equipment storage area and lounge for field crews.[15]

Lubbock County Judge Rod Shaw and Alton Brazell, Lubbock County Commissioner for Precinct 4, which encompasses the Lubbock Lake Landmark, also sat on the LLSDC board. Nash worked with them to provide county services to the Landmark. Johnson says that to protect the site from accidental damage, Brazell always sought her input and approval before work began. The county dug a large holding pit through the dry reservoir floor, which was already in bedrock, to create a matrix pond for

Figure 15. After the lake dried out, the county dug the Landmark's original matrix pond into its bottom. The county brought in water and periodically cleaned the pond to remove buildup. (Courtesy of Lubbock Lake Landmark)

washing sediment samples from the excavation areas. Breaking through bedrock requires heavy machinery, which the Landmark could never afford.[16]

Brazell, who shared Nash's love of history (primarily agricultural history), would remain involved with the Landmark personally and as a commissioner until his retirement in 1994. His precinct's road crews maintained the staff access roads around the site's perimeter, repaired the old walls surrounding the reservoir, kept the grounds, hauled trash, and improved the drainage system. The county also delivered a surplus trailer for crew housing and an evaporative cooler for the kitchen, which Brazell had installed himself. Johnson notes that he could often be seen running the heavy equipment used at the site.

Providing housing and food for the field crews was vital, as many came from across the United States and other countries.

Figure 16. Alton Brazell, Lubbock County Commissioner, getting ready to install the Landmark's first evaporative cooler in the field kitchen. (Courtesy of Lubbock Lake Landmark)

Today's crews are still multinational. Volunteers lived in tents at the Landmark until portable buildings were built after the 2024 season. Crews at regional sites live in landowner-provided shelters.

In 1977, LLSDC received a National Endowment for the Humanities grant for a master plan for the site. As work on the plan continued, Brazell convinced the commissioners court to fund a prefabricated building outside the fenced 29-acre site as an interpretive center. A 1979 contract between the university and LLSDC called for the creation of exhibits. Owing to a variety of delays, the building sat empty for years. In the late 1980s, Brazell convinced the commissioners court to move the structure inside the fencing and renovate it as a lab for the field site, a function it still serves.[17]

In 1980, Lauro Cavazos became Texas Tech University's president. Nash lost no time in giving Cavazos a tour of the Landmark. Cavazos, who'd earned his bachelor's and master's degrees in zoology from Texas Technological College before earning a doctorate in physiology from Iowa State University, appreciated the importance of the Landmark. Within a year, a series of discussions led to a letter of agreement between the LLSDC and the university. In the letter, the university agreed, through the museum, to take full responsibility for developing the Landmark as an educational and research facility. In her 2016 book chapter, Johnson wrote: "This responsibility was viewed as encompassing the educational, scientific, research, and cultural aspects of the site, with an interpretive program . . . emphasizing the scientific and cultural significance of the site."[18]

The agreement marked the first time the university had explicitly committed to funding the Landmark operations. Negotiations to transfer the LLSDC's lease on the twenty-nine acres to the university were successful, giving the university, under the auspices of the museum, control of the acres that made up the Lubbock Lake Landmark. Governance on the entire 336 acres would not be resolved for years.[19]

The lease transfer marked the end of the LLSDC, with the organization no longer seeing its efforts as necessary to protect and develop the Landmark. However, before the LLSDC dissolved, its members negotiated with the National Endowment for the Humanities to transfer the master planning grant fund to the university, allowing Johnson to lead the planning. She and her team produced a three-volume plan. Including a philosophical and programmatic discussion of the Landmark's significance, the plan assessed past decisions' negative impacts on the land, outlined needed resources, and provided options to prevent further damage. The plan called for controlled public access to the site for year-round educational and research activities—and, most significantly, for long-term protection of the entire 336-acre site.[20]

With events headed toward a positive trajectory, an unexpected crisis occurred in the form of a rising water table. Although the reservoir had been dry for decades, water would pool in the matrix pond over the summer, which helped with sediment processing, but would dry up during the winter. But in 1979, the water table had risen to the point where water remained in low spots of the reservoir floor year-round. By 1984, the reservoir was flooded, stopping new excavations and forcing the abandonment of existing work, a problem that continues today. The issue affected much more than the Landmark; it also flooded city parks and some Texas Tech buildings. The water was only about six inches below the playing field at the football stadium. The rise in the water table was not, Johnson says, owing to a rise in the Ogallala Aquifer, as officials hoped, but was human-made when the water from agricultural and urban runoff could not get through the hard caliche caprock above the Ogallala Aquifer, a perched water table formed because the water had no place to go but up. Johnson notes that, in addition, the way playas were used across the region contributed to the rising perched water table. Playas were drained for several reasons, including urban development and roadway construction.

A POWERFUL ALLY

Nash's continuing efforts and friendships with local leaders resulted in a significant turning point for the Landmark in the mid-1980s when he brought then-State Senator John Montford to the table. Montford, whose district included Lubbock, was an influential legislator. He chaired the State Affairs Committee, which deals with major legislative issues, for two legislative sessions, and the powerful Senate Finance Committee, which writes the state budget, for four sessions. The State Affairs and Finance Committees were the two most powerful State Senate committees then.[21]

Like Nash and Brazell, Montford was a history buff. He had an extensive collection of spurs that he eventually donated to the National Ranching Heritage Center at Texas Tech. He and Nash were members of the Western history enthusiasts' organization, the Westerners. "It was through the Westerners that I met Bob not long after I moved to Lubbock," Montford says. "He was highly respected in Lubbock."

Montford says that while Nash talked up the Landmark, Johnson truly got him interested. "Eileen convinced me the Landmark was a unique archaeological site, not just for Texas but internationally," he says. "I was just astonished at everything that had existed out there."

Montford set about figuring out funding that would allow the development of the Landmark as outlined in the 1980 Letter of Agreement between the city and the university and that would meet the goals of the master plan. With Montford engaged, so too was one of his top aides, Morris Wilkes. Wilkes served as Montford's chief of staff and later Chief Clerk of the State Affairs Committee while Montford was chair. Johnson recalls that Wilkes acted as a liaison between various agencies and stakeholders and Montford, a role that would increase after Nash died in 1993. Johnson says Wilkes was active behind the scenes in promotion and fundraising, assuming Nash's role as "prodder, facilitator, and advocate." Montford picked up where Nash left off as a public spokesperson and visionary for the Landmark.

Wilkes was not only on Montford's staff but also a long-time friend of Nash's. He had worked with Nash in radio in the mid-1970s. Like Judge Parrish, Wilkes remembers Nash telling stories about regional history and the Landmark. "During the summer, Bob and I would take a pickup that was kept at the radio transmitter and load it up with watermelons to take to the field crews," he says.

A NEW PARTNERSHIP

One of Montford's first efforts on the Landmark's behalf was to bring Governor Mark White for a site tour in early 1986. By midyear, Montford had talked to the Texas Parks and Wildlife Department's (TPWD) executive director about the continuing water table issue and constructing an interpretive building at the Landmark. Montford brought TPWD and the university together, and in 1987, they formed a first-of-its-kind partnership between state agencies that resulted in the development and preservation of the Landmark. The agreement made the Landmark part of Texas's park system. The university, through the museum, took responsibility for all research and educational activities. TPWD would provide funding for building construction, landscaping, and security. In his words, Montford's goal was to make the Landmark "a place that was an attention-getter, that would draw people in to recognize what we had here."

Why a partnership with TPWD? Montford explains it was simply the quickest way to fund the construction necessary to turn the site into what he envisioned. Andrew Sansom, who became TPWD executive director in 1990, says that with Montford leading the partnership effort, it wasn't hard to convince the agency to agree. "We had maybe two or three legislators who consistently supported the Department," Sansom says. "We knew that Montford would always come through for us."

Montford also facilitated a new lease between the city and TPWD, replacing the lease held by LLSDC. The fifty-year lease covered the entire tract designated the Lubbock Lake National Historic and State Archeological Landmark. In 1992, a land swap between the City of Lubbock and TPWD gave the state agency the Landmark area, while the city would acquire the 542-acre Mackenzie State Park, making Mackenzie a city park. The exchange settled the question of who would control all 300-plus acres of the Landmark for the moment.[22]

A formal dedication of the Landmark as part of the state park system came in 1987, drawing statewide media and a range of elected and community leaders. The following year, TPWD set about trying to resolve the water table issue. TPWD, working with the university's Water Resources Center, devised a plan to drill

Figure 17. At the ribbon cutting of the new Robert R. "Bob" Nash Interpretive Center are (L–R): John Montford; Bob Nash; Lubbock mayor B. C. McMinn; Curry Holden; and Texas Tech President Robert Lawless. (Courtesy of the Southwest Collection/Special Collections Library, Texas Tech University)

about thirty wells around the reservoir. The wells would connect to a pipeline carrying the water further down the canyon. Although the project looked like it would function, it was expensive and would have significantly damaged archaeological deposits. The water table issues remain unresolved.

A TPWD master plan built upon the one created by Johnson was released in January 1988. It laid out proposals for constructing self-guided visitor trails with interpretive signage, a research building, and an interpretive center to provide exhibit space and serve as a headquarters building, along with parking and a picnic area.

Texas Tech was to fund, develop, and staff continuing archaeological research, create educational programs for public schools and other interested groups, organize a nonprofit support group to assist in educational programming, operate a gift shop, and produce publications and promotional materials. The museum and TPWD's exhibits team would work together to create exhibits in the interpretive center to educate the public about the archaeological work.[23]

With facility construction well underway, a practical matter surfaced. Wilkes remembers that he and then-TPWD director Dicky Travis were talking with State Senator Montford in the latter's office late one day during the 1987 legislative session when the question of what to name the interpretive center came up. "I asked the question, and Dicky said he hadn't thought about it," Wilkes recalls. "I remember that John specifically looked at me and asked what I thought. I said it should be named for Bob Nash. Without him, we probably wouldn't be having this conversation."

Nash was surprised later that year when officials announced at a news conference that the Interpretive Center would bear his name. A resolution approved by the City Council would honor Nash and commend him for his "tireless dedication and years of service to bring this project to fruition."

A 1988 groundbreaking ceremony brought another round of statewide attention. On December 26, 1988, a *Dallas Morning News* newspaper headline noted experts were saying, "Lubbock may be one of [the] New World's oldest towns." The story described the event as a "very gentle groundbreaking" in reference to the fragile artifacts just below ground level. The reporter noted that "designing the park buildings so that no archaeological treasures are destroyed or damaged has required stringent restrictions on construction." Johnson told the reporter that a few places within the park were not archaeologically significant, such as the Blanco bedrock, where the interpretive center and Quaternary Research Center building would stand.[24]

A year-long grand opening began on October 14, 1990. The opening ceremony drew around five thousand people, including dignitaries and media from across the state, to celebrate the new facilities and fifty years of research and educational activities at the Landmark. The interpretive center finally became a reality, as did a new laboratory to support research efforts. Crews also completed hiking trails, a picnic area, and a parking lot. International scholars attended a symposium designed to put the Landmark's research into a global perspective. The symposium resulted in a 1995 peer-reviewed publication, *Ancient Peoples and Landscapes*, attesting to the worldwide significance of the Landmark itself and the research contributions there.

LIFE-SIZE EDUCATION

The grand opening closed one chapter in the Landmark's history and opened a new round of political and development activity. Montford lost no time in organizing the Lubbock Lake Site Foundation to raise funds. Nash, Montford, Brazell, Johnson, then-museum director Gary Edson, Wilkes, and other influential community members were on the nonprofit's board.

First on the agenda was raising funds to create and install four life-size sculptures of extinct animals from the end of the Pleistocene that researchers had discovered at the Landmark. A Columbian mammoth and her calf, a short-faced bear, an ancient bison, and a giant pampathere (a distant cousin of today's much smaller armadillo), are located around the interpretive center. Johnson notes that the statues serve not only as art but also as educational and promotional pieces.[25]

The short-faced bear captured Montford's fancy. "Eileen told me he was a mean critter," he says. Montford wanted the sculpture to depict the bear standing erect with its paw raised, believing it would attract the attention of cars passing by on nearby Loop 289. Researchers estimate the creature would have stood up to 12 feet tall on its hind legs. "It was a big animal, and I thought maybe we could use that to promote the Landmark." Montford lost that battle, and the bear stands on all four paws.

As artists completed each sculpture, public events drew hundreds of people to the Landmark, with each event focused on

Figure 18. The short-faced bear sculpture was John Montford's favorite creature. He examines the giant bear's tooth during the dedication of the sculpture. (Courtesy of the Southwest Collection/Special Collections Library, Texas Tech University)

the community in general and schoolchildren in particular. Since Curry Holden first explored the reservoir area, the Landmark has prioritized educational activities, and Johnson has always carried that emphasis forward. She is grateful that Montford understood the importance of promoting both education and research.

Second, but perhaps most important, the foundation focused on establishing a secure funding base for research. Montford worked in the state legislature to obtain stable funding and increased state assistance for the Landmark. In 1991, he procured an appropriation through the museum for two additional Landmark employees. In 1995, Montford's work resulted in legislative approval for increased state financial support to cover operational costs by creating an appropriations line for the Landmark that is separate from the museum's appropriation. Montford, Edson, and Johnson also sought to create an endowment for long-term research and educational support, a goal they did not achieve. However, an endowment from a local foundation continues to provide much-needed research funding today.[26]

A NEW ERA

The Landmark lost its original champions when Holden and Nash died in 1993. Montford left the State Senate in 1996 to become the first chancellor of the Texas Tech University System. His focus shifted to developing the newly established system and increasing the university's endowment. Wilkes redirected his attention to national and statewide politics. The Lubbock Lake Site Foundation was dissolved.

The Landmark–TPWD partnership produced much-needed facilities funded by the agency. Andrew Sansom says that while the agreement with the Landmark was unique, it made sense because the agency managed all the state's historic sites then. The Landmark was, however, a very different sort of place from others under the TPWD umbrella. Sansom notes that only Dinosaur Valley State Park—near Glen Rose, about an hour southwest of Fort Worth—was remotely similar to the Landmark, but explains that the Fort Worth site is a traditional state park with camping and recreational activities, including allowing visitors to walk in the dinosaur tracks along the Paluxy Riverbed.

While Johnson observers that governance issues arose, Sansom does not recall significant problems with the partnership. A Texas Tech graduate and environmental conservationist, he would go on to serve as a research professor of geography and executive director of Texas State University's Meadows Center for Water and the Environment before he retired. He asserts that the Landmark is a globally significant site and feels privileged to have participated in its management and development.

"The partnership was especially significant to me personally," he explains. "I did not detect any problems, and given that I was really interested in that site, I think any significant issues would have been brought to my attention."

From Johnson's point of view, a significant issue centered on the fact that each agency was accustomed to having sole responsibility for managing and programming its properties. While the first-of-its-kind agreement outlined responsibilities for each agency, Johnson saw an immediate area of concern. She says each agency had its own procedures, management styles, and personnel who had to figure out how to work together despite differing perspectives on governance and stewardship.

Although the two agencies agreed on many things, Johnson explains that her issue came down to TPWD's focus on revenue and recreation versus the museum's mission of education and research. Although TPWD receives state funding, it counts on income from recreational parks to meet its budget. "In the eyes of TPWD, the most successful parks were the revenue-generating recreational parks; thus, the least successful parks were the nonrecreational historic parks," Johnson says. "From our point of view, the museum and the university expended funds to provide learning experiences and research; we never expected activities to yield a profit."

The Landmark had never charged an entry fee. Johnson says that as far back as Curry Holden's time, the belief was that people had already paid through their tax dollars and contributions. Johnson says that initially, there was a verbal agreement between the agencies that park entry would remain free. Eventually, under a mandate to generate more revenues, that agreement collapsed, and an entry fee of $1 for adults and 50 cents for children (children under six were admitted free) was instituted.

By the late 1990s, TPWD was under pressure due to the poor condition of some historical parks. In 1999, the Texas Legislature allowed TPWD to divest itself of historic parks so long as a suitable partner was available to assume control. While Montford had left the State Senate for the Texas Tech University System, he played a significant role in assuring that the transfer of the Landmark from TPWD to the Museum of Texas Tech University was the first to take place under the new legislation.[27]

Two years of funding would be available under the terms of the transfer. Per the separation negotiations, a maintenance yard and a park manager's house would stay with TPWD as part of its regional headquarters adjacent to the Landmark. The Landmark transfer was set for September 1, 2000, but the legal transfer took a few more months to finalize. By early 2001, the Landmark was governed by the Museum of Texas Tech University, but with a director and operational staff separate from the museum's administration.[28]

The newfound control brought both successes and continuing problems. With many objectives met and problems resolved, new iterations of some issues persist. Johnson says the raised water table remains problematic and is particularly destructive to organic materials, which form a significant part of the Landmark's record. Fundraising is a continuing necessity. Staffing reductions from past legislative budget cuts have meant fewer people to maintain and operate the site.

Yet there have been many positives since Johnson arrived at the Landmark. Annual summer fieldwork provides year-round research opportunities. The site is protected from vandalism and unauthorized digging. Substantial damage, resulting from city-sanctioned use of the northern part of the site by off-road vehicles and motorcycles, has stopped. The Nash Interpretive Center is open year-round, offering exhibitions, educational activities, and tours. The entry fee was eliminated.

The Landmark continues to revitalize parts of the prairie that are not under excavation, restoring those acres to a natural grassland, as it might have been in the late 1800s. Staff built a new ADA-compliant wildflower trail. It took seven years to construct, using a method that did not require digging into the soil. Crews have removed invasive plants, allowing native plants and wildflowers to return. The mixed-grass prairie now provides wildlife more protection, and biodiversity has increased. The Texas Horned Lizard (best known as a horny toad), listed as a threatened species in the state, has made a comeback at the Landmark. Deer, squirrels, grey and red foxes, and coyotes have also moved into the prairie.

The Landmark's trails, featuring informational signs, are popular with the community as places to walk and bicycle. The site is beautiful in the spring and summer when covered in wildflowers, provided rains have come at the right times. Public education and outreach efforts are vital to the Landmark's continued community support.

Johnson's long tenure and interdisciplinary approach to research have been central to the Landmark's survival. The Landmark and the scientist are a perfect fit. Johnson's focus on the entire Quaternary Period has produced impressive discoveries at the Landmark and in several locations around Lubbock and the region. Her approach has also dictated how crews conduct excavations. When a fossil or artifact is discovered, the excavation crews collect everything in the dirt around it to determine what else can be learned.

CHAPTER 4

DIGGING UP HISTORY

Anthropology, archaeology, and paleontology are the primary sciences scientists use to uncover our ancient past. Anthropology studies the origins and development of human cultures. Archaeology, a subdiscipline of anthropology, examines material culture from the past and present to understand human behavior. On the other hand, paleontology studies ancient animals, including dinosaurs, to decipher how creatures evolved and what the Earth was like millions of years ago. It is important to note that archaeology and paleontology also study remains that are only a few thousand years old, such as the plants and animals that lived in and around the Landmark and were used by the people who lived around the lake.

While all three disciplines overlap, Johnson argues that the work at the Landmark is much broader than textbook definitions. She believes an interdisciplinary approach is necessary to understand and interpret whatever evidence is uncovered. Johnson and the people who work at the Landmark are conversant in all three areas, as well as stratigraphy, geology, geochronology, geology, paleobotany, biology, and several other scientific fields.

Although Johnson's degrees are in anthropology and zoology, she does not consider herself an anthropologist. "If anything, I consider myself a Quaternary scientist because of the interdisciplinary nature of our work," she explains. "We are focused on the last 2.6 million years, most of which have nothing to do with people. Our work spans a wide range of disciplines, and this is a perspective I share with everyone I work with, especially students."

Johnson believes that such perspectival breadth is necessary to comprehend the linkage between the period's animals, plants, humans, climate, and environment. Her focus on interdisciplinary science began early in a seventh-grade science class. She recollects that the teacher talked about several types of science, ranging from volcanology and anthropology to ornithology and various biological and natural sciences. "I remember having the feeling [then] that I knew what I wanted to do with my life," she recalls with a smile.

That early realization carried over to her graduate studies. While working on her master's degree, she didn't limit herself to the traditional anthropology courses. "I also took geology, paleontology, and biology classes," she says. "People told me I was wasting my time and should only take anthropology courses. Now, a lot of universities are stressing interdisciplinary courses, trying to remove the silos that I broke down a long time ago."

The excavations at the Lubbock Lake Landmark and its regional sites provide a glimpse into the world spanning the past 2.6 million years. No evidence of dinosaurs has ever been found at the Landmark or across the Llano Estacado. The giant T-Rex and its like died out 65 million years ago. The Ogallala Formation and the Llano Estacado did not begin forming until 23 million years ago. If there are dinosaur fossils at the Landmark, they would be buried beneath the bedrock. Digging beyond that extremely hard surface would be expensive and time-consuming, with no guarantee that researchers would find anything.

As with any scientific endeavor, the work at the Landmark begins with a research question. The question may become more focused depending on discoveries, changing knowledge, and a particular researcher's interest. Some scientists start with specific questions about a particular period or artifact that are answerable in one or two field seasons. Others, like Johnson, begin with a question requiring a larger answer. For her, the question was: How have humans adapted to ecological change on the Llano Estacado over the past 13,000 years? "Despite more than fifty years of work, we're still trying to answer the original question from 1973," she says. "The goal of the first field seasons was to discern the various cultures and environments that existed in the area for the past 13,000 years and how they relate to each other." Johnson knows she will never answer her big question, but the Landmark work contributes to reconstructing long-extinct cultures. Discoveries can produce multiple subquestions, such as those that guide Stance Hurst and Doug Cunningham's work on how the Apache and Comanche tribes used the land. The pair also focuses on buffalo hunters and early ranch history.

Hurst, the Landmark's research field manager and assistant professor in the TTU Heritage and Museum Sciences master's degree program, has a personal connection to the Landmark. He joined the team full-time in 2007 after completing his doctorate at the University of Oklahoma, but his journey with the Landmark began in 1995 when he spent a summer as a field crew volunteer. "It was my first fieldwork experience, and it fit me," he reflects.

While he worked on archaeological sites in Idaho and Oklahoma, serving in a museum setting was new. "We do traditional archaeology here, but because we have a museum focus, everything that's done here is much more detailed," he explains. "Other archaeological sites may sample a site. But we are careful to document everything from when we begin digging until the bone or artifact ends up in the museum."

Figure 19. The sediment surrounding the bone or artifact is placed into large plastic mesh bags, known as matrix bags, for examination at the Quaternary Research Laboratory at the Landmark. The excavation is located in Area 8, near the Singer store. (Courtesy of Lubbock Lake Landmark; Jim Gannaway, photographer)

Hurst points out that Johnson's broad range of interests makes the Landmark excavations different from most. The Landmark crews are not just looking to identify an object and determine its age; they are also looking at the conditions in which it existed. What was the temperature range? What did the landscape look like? Was it dry or wet? To get to those answers, the crews carefully gather all the soil, sediment, and other materials that may be found with an artifact or fossil, resulting in hundreds of range bags and buckets of material coming into the Landmark's laboratory by the end of each dig season.

The primary targets of excavations are artifacts, fossils, and sediments. Artifacts are human-made or -modified objects, such as tools or weapons fashioned from bone or stone. From artifacts, scientists can learn about the diet, tools, weapons, and living conditions of the people who made them. Because stone survives much longer than bone-made artifacts, stone objects, called lithics, a subcategory of artifacts, are what are most often found. Fossils are the bones and tracks that tell archaeologists about ancient creatures. Sediment is loose, unconsolidated material (basically dirt), made up of rock fragments, minerals, and organic remains that is transported by wind, water, or ice and deposited in a new location. All play a role in revealing how the environment has changed and how animals and humans lived and developed over the millennia. Landmark crews have uncovered animal bones dating back millions of years and human artifacts dating back about 13,000 years.

The first step in any excavation is simply identifying where to dig. In the region surrounding Lubbock, a landowner or citizen will sometimes pick up what they believe to be a bone or artifact and contact the Landmark. From there, the scientists determine if the find is worth pursuing. A citizen's comment can engender a new location for exploration, such as a location that opened in 2024 east of Lubbock. The new exploration was sparked when a man visiting the Landmark saw an exhibit about dugouts on the South Plains and told the Landmark folks he thought that he had a dugout on his property.

Deciding to excavate is a significant commitment. The Landmark depends heavily on volunteers to staff its crews. With numerous active dig sites throughout the region and excavations at the Landmark, resources get stretched thin each year. With the short dig season (from May to August), crews may spend only a few days or weeks at any one site each year before moving on. However, the crews visit many of the same sites annually, investing significant time and resources over the years.

Several things happen before a site is measured into one-meter-by-one-meter squares and the delicate digging begins. Every excavation starts with a variety of tests and surveys, depending on the site. Typically, crews will do a pedestrian or metal detector survey where the crew walks across the area in a straight line to see if anything is visible on the surface. Bones and artifacts may stick out of the sides of draws or be only partially buried in a flat surface, thanks to erosion. Rain and wind can eventually expose objects hidden for thousands of years.

Any bones or artifacts on top of the ground or protruding from the walls of a draw are marked and photographed or sketched in situ, and measurements of the exact location are taken to establish the relationship between found bones or artifacts and the landscape before excavation begins. Crews also periodically walk around every site to see if anything new, such as a bone or artifact, has eroded to the surface, which is how new places to dig are often discovered.[1]

A more extreme way to explore a site involves cutting a short trench with a small backhoe. The crew is not necessarily looking only for objects; they are also looking at the stratigraphy, which will give them an idea of the area's age. Many of the sites the Landmark crews are working on have been in the field season rotation for several years, meaning that scientists already have radiocarbon dates, a firm understanding of the stratigraphy, and an idea of what more the sites could yield.

Radiocarbon dating, which the Landmark uses extensively, allows scientists to determine the age of organic material as far back as 60,000 years. A permanent exhibit at the Landmark's Nash Interpretive Center explains the process. Carbon-12 is an essential element of life for all plants and animals, along with a tiny amount of radioactive Carbon-14. Carbon-14 decays into Carbon-12 as radioactivity is lost, a process that stops at death. By measuring the amount of Carbon-14, scientists can determine how long the plant or animal has been dead. That date gives the Landmark researchers the approximate age of the stratum, or layers, of the earth in which the material was found. Any artifacts found with the sample are of the same approximate age.[2]

Crews also map a site using several methods—including traditional land surveying techniques with a scope and measuring

Figure 20. At every excavation site, researchers take careful measurements to document the exact location of each bone or artifact. (Courtesy of Lubbock Lake Landmark; Jim Gannaway, photographer)

rod, similar to those used by road-building crews—to determine a landscape's horizontal coordinates. A two-person crew walks around the site, one taking measurements and the other writing down the coordinates. A GPS (Global Positioning System) maps the location of objects and places in the field. The GPS data, as Stance Hurst describes, is incorporated into GIS software (Geographical Information Systems). "The GIS manages all of our geographic data. It allows us to examine the interrelationship between our mapped objects and places in relationship to other sources of geographic information, such as geological maps and hydrological maps."

Another indication of how much technology has changed over the years is that Hurst has begun using drones in the mapping process to enhance data collection. "With traditional surveying methods, you can gather hundreds of data points," Hurst explains. "With the GPS we use, we can record thousands of data points as we walk the site. But a drone can collect millions of points in minutes." Surveys are not a one-time occurrence. Sites, including the Landmark, are resurveyed periodically.

Johnson notes that her first work at the Landmark in the early 1970s focused on chronology and identifying what groups lived there. "Now we're looking back with what we've learned through the decades to see how a specific group may have defined its territory," she says. "We want to know how they defended their territory; did they allow other groups into their area? How did the changing climate affect the ecosystem? Hopefully, the questions we are asking now are more incisive because we are building on that initial level of questions."

When evaluating dig sites, scientists also utilize factors such as distance from water, soil type, and other factors that may have influenced where humans settled, even briefly, or animals gathered. They also conduct archival research, looking at existing databases, historical documents, previous research, if any, or oral histories to learn more about the site. Regardless of whether a site is deemed significant enough to excavate, researchers write a site report to document the work and any objects found.[3]

Archaeology is, by nature, a destructive endeavor. When a bone or artifact is removed, the site can never be reconfigured exactly as it was. Although a site may encompass a large area, such as a portion of a ranch or an ancient draw, excavations are typically a small sample of the site. An entire site is almost never fully excavated.[4]

Once archaeologists decide to dig, the excavations are usually conducted in a series of one-meter-by-one-meter squares. Crews then take more measurements, determining each square's relationship to the general area and other dig areas.

The types of tools archaeologists employ may vary with the sites under excavation. Still, for most excavations, the toolkit typically contains a wooden pick, similar to a long toothpick, to carefully remove the dirt around a fragile artifact. Brushes of varying sizes lightly sweep away loose dirt. A professional trowel is also used in the excavation. While it may resemble one used in a garden or to apply mortar, it is much smaller across its base, often an inch or less. Metric tape measures are necessary to determine the artifact's exact location accurately.[5]

Throughout the process, crews meticulously document everything. Before removing anything, they create a site grid to provide a fixed reference point for all measurements. A grid divided into frames representing the one-meter-by-one-meter squares is then overlayed onto the site grid. Each square is measured and assigned a number. This procedure produces a scale-drawing map that allows researchers to pinpoint the exact locations of every bone and artifact at the site and to show the relationships between artifacts recovered in situ.

This documentation also comes into play when dating artifacts and when crews return to a site that has not been looked at in years. During the 2023 dig season, crews worked next to excavation units from the 1973 and 1974 seasons, the first done at the Landmark after Johnson arrived. The question is, what else can we learn based on the knowledge gained over the decades? Crews know what was found then: a hearth, evidence of bison butchering, stone tools, and pottery. "I know exactly where those objects were found," says Lila Jones, a research associate and the crew chief at the Landmark and other regional digs. "Because I know the exact measurements of the early squares, I can correlate that data with what we find now. There was a campfire in the older excavation; maybe we will find another one in our new block or something completely different. Our documentation gives us a better idea of how the space was used."

Crews collect everything found around an object. Answers, or even new questions, can be unearthed in the dirt that surrounds what is found in an excavation. However, what researchers are looking at is not just dirt; it is the sediments, soils, and other organic materials that surround a fossil or artifact that are important.

Figure 21. Matrix bags contain dirt and other materials that might surround artifacts and bones. Once bags are brought to the Quaternary Research Laboratory, the contents are washed through fine mesh screens. (Courtesy of Lubbock Lake Landmark; Kippra D. Hopper, photographer)

Jones and Katie Holt, the former associate collections manager for the Quaternary Research Center (QRC), begin a different research phase once a dig site is closed for the season. The associate collections manager's job includes performing an initial inventory when things come in from the field to ensure everything is accounted for based on records from each field site. Jones processes samples from the field and recommends what should be sent for radiocarbon dating. As material comes in, she assigns a number to all the samples collected in the field. Jones explains that researchers look at organic matter as soon as possible so that it does not degrade further and hamper any possible radiocarbon testing.

The QRC receives hundreds of matrix bags, boxes, and buckets of all that is collected in the field each year. Everything arrives in individual bags, accompanied by an object tag that provides provenance information and a basic identification of the object. The initial identifications made in the field are confirmed or corrected once in the lab. Two tags are created for each matrix bag, one placed inside the bag and the other attached to the outside.

Next in the process, both tags are retrieved and examined to ensure they match. If those tags are lost, then all the information and significance of an object is lost or diminished. Then, the staff washes everything upon arrival in the lab. All the soil and sediment are washed through a fine two-layer mesh screen to catch any tiny artifacts, bones, seeds, or snail shells that may be hidden. Once the material dries, the QRC crew sorts what is caught in the screen to see if there are tiny pieces of bone, rock flakes, or pieces of metal and ceramic. Their primary focus is on recovering microvertebrates, botanicals, and other biological materials. Graduate and undergraduate students as well as volunteers work in the QRC. Staff members provide examples for comparison. It helps those washing materials to establish whether they have found a regular caliche rock or a possible hearth stone. It's a tedious yet important job. Small bits of bones, a lizard skeleton, or a tiny shell help Landmark researchers learn, for example, what the environment was like. Sometimes, a bit of organic material or charcoal is found, and Jones determines whether it should be sent for further identification or dating. Any bones or other artifacts found during the season are cleaned and given to Jones or Johnson for final identification. A spreadsheet is kept, and each item is updated for the database for each site. Once everything is processed and identified, a final inventory is taken, and packages are sent to the museum for permanent storage.

The sheer amount of material that comes through the QRC—from tiny seeds to bullet casings to large animal bones—means the processing may continue until the next summer's field season. A single field season can generate thousands of objects, each with its own catalog number. All field records, including site survey forms, excavation field notes, sketches, maps, photographs, and specimen provenance records, are filed for future reference. Provenance denotes the actual place where an object is found. Some records are used in a report filed with the Texas Historical Commission (THC) every five years to retain the Landmark's state permit. The report is necessary for any work done on state land. The THC administers the Antiquities Code of Texas and the State Antiquities Landmark program. As a state archaeological landmark, the Lubbock Lake Landmark is subject to the laws of the state of Texas.

While the dig season begins in May, preparations start months before. By January of each year, the Landmark begins soliciting volunteers to work in the field and the laboratory. The volunteers, who come from all over the world, are the lifeblood of the research program. The number of volunteers determines how much is accomplished each summer. The numbers vary yearly, but the program requires a lot of human resources to do the current research.

Whether volunteer or paid Landmark staff, working on an excavation site in the scorching South Plains heat is not a vacation; the labor is hard. Sitting on the ground and digging around an object may not sound strenuous. However, there is little shade, and temperatures can reach near and above 100 degrees Fahrenheit. The workday runs from 6:45 a.m. to 4:30 p.m.; all crew members also take on cooking, kitchen duties, and other camp chores. "We run primarily on volunteer help, so we're focused on the summer when most people can get away from jobs and their normal lives," Johnson explains. "We differ from many programs because we are out for four months straight. A lot of places are out only four to six weeks."

Living conditions at the regional sites vary, but most are less than luxurious. At the Post, Texas, location, the landowner provides crews with a ranch outbuilding, basically a two-room barn that has electricity and is air-conditioned. There are clean and maintained outhouses. Water is trucked in, and a system is set up for showers and a sink. There is no internet service, and cell phone reception is iffy. At the Landmark, crews lived in large tents with wood floors until 2025, when new air-conditioned cabins were installed. There is a kitchen and recreation room.

Volunteers get hands-on experience in an authentic archaeological dig in exchange for their time and labor. Some are so enthusiastic about the experience that they change their college majors. Others get to take a trip down memory lane. When Grayson Kirk Meade, a retired lawyer living in New York State, decided on a whim to volunteer in the summer of 2019, he was coming home.

"I was fiddling around on the internet one day and looked at the Landmark website because I had a long-ago connection to here," he explained when interviewed at the Landmark in 2019. Meade grew up around the early excavations of Lubbock Lake. His father, Grayson E. Meade, was a geology professor at what was then Texas Technological College. Meade Sr. and Glen Evans conducted the first fieldwork inside the now-dry reservoir's banks in the early 1950s. Their work established Lubbock Lake as a significant early human site.

Meade remembers coming to the dry reservoir, wandering around, and looking for arrowheads while his dad worked. "I had no idea just how significant the place has become," he says, noting much of that has to do with Dr. Johnson. "I know my dad and Glen Evans are sort of idolized here. It is weird to think this is my dad, who had such a significant influence." Meade did not get to work in the exact spots his dad had worked, but he was able to get close.

During the 2023 summer season, Markus Crawford was in his third year as a volunteer. "I saw the volunteer program and thought that it might be interesting," he said in an interview as

Figure 22. Markus Crawford (left) and Cameron Ditore work on an excavation in Mustang Draw on the Yater Ranch in 2023. (Courtesy of Lubbock Lake Landmark; Jim Gannaway, photographer)

he continued scraping around one last piece of bone to remove it before the site was closed for the summer. "I worked the first summer session and signed up for the second."

The draw for Crawford is the interdisciplinary nature of the Landmark's program, as it is for Cameron Ditore, who entered the Heritage and Museum Sciences master's program in 2024. "By the second or third weekend of that session, I called my academic advisor and said I was changing my major to anthropology," he says. "I suspect I am not the only student who has done that."

As he answers questions, Ditore works in the next square over from Crawford to remove another bone. The two were at a site on Yater Ranch near Stanton, Texas. Their work typifies the patience needed during excavations. Ditore uses small paintbrushes to remove the dirt around the bone. "We try to get as much dirt off as possible. The bone is fragile; you could almost just breathe on it, and some of it would come off," he says.

Both men carefully filled out all the documentation and mapped where they found artifacts. They were working on what was a Comanche campsite. "You have to have the spatial location of everything to understand what they were doing at the campsite," Ditore explains. "Concentrations of bones can mean they were butchering bison, or lithics indicate they were making stone tools. You have to know where everything is to make sense of the area. It is sort of like a forensic crime scene."

Crawford and Ditore were in an area where a small glass bead was found in 2022. The bead was discovered above Apache-era material and is believed to be Comanche. Ditore and Lila Jones presented their research on the bead at a conference in late 2022, a boost for their academic and professional resumes. In addition to finding bones, they keep an eye out for more beads. Ditore says he and Jones believe the bead was probably traded between the Spaniards and the Comanches. "We know there was another similar bead found in a mission in California and another in Central America that was perhaps from a necklace passed down through generations," he theorizes. "So, wherever the bead came from, it was either from a raid by the Comanches or traded for something else." While the Comanche are perhaps best known as exceptional riders and fierce warriors, they were also skilled traders, allowing them to control much of the commerce of the Southern Plains.

The bead's history is just one of the many mysteries researchers try to solve. Is it from the mid-1700s, a time that would line up with other things at this dig site, or is it possibly from the 1800s and representative of another, later occupation of which crews have found little evidence? As with the dozens of questions that come up each summer excavation, perhaps future dig seasons will provide an answer.

Even though the Lubbock Lake area is the mother ship for the Landmark's archaeological research, regional excavations, such as the Stanton site, undertaken by Landmark crews, add to the understanding of the area's natural history. These regional sites have unearthed signs of animals much older than those found at the Landmark. There are also signs of human and animal occupation in the regional digs that do not appear around Lubbock Lake. However, as Johnson notes, almost every discovery relates in some way to the Landmark, supplementing and reinforcing the record there.

CHAPTER 5

LIONS, BUFFALO HUNTERS, AND A GIANT TORTOISE

Each of the Lubbock Lake Landmark's regional sites has yielded a variety of fossils millions of years old, as well as human-made artifacts dating back at least 13,000 years. Each discovery contributes to our understanding of the animals and humans who once inhabited the Llano Estacado and the surrounding Caprock region. Some sites have numerous excavation areas.

The regional sites are only accessible thanks to the generosity of the landowners, who are happy to have the Landmark crews and excited about what the latter are doing on their property. "They visit and see what we're up to," Hurst says. "They want to learn about the past as it applies to their land. The historical archaeology we do is often their history."

Hurst points to Clovis and Folsom materials found on the U Lazy S Ranch near Post, Texas, as an example of how regional discoveries enhance understanding of those at the Landmark. Researchers found a Clovis projectile point in 2014, dating to about 13,300 years ago. Paleoindian sites are rare on the Southern High Plains, Hurst notes. "Finding additional Clovis material provides important research opportunities to examine how these early hunter-gatherer groups interacted with a dramatically different landscape," he says. The Folsom-age material is likely from a bison kill.

Both finds come from the Spring Creek area of the U Lazy S Ranch. John Moretti, former research aide and senior crew chief, describes the U Lazy S land as yielding some of the richest Pleistocene animal discoveries outside the Lubbock Lake Landmark. However, the ranch holds much more than fossils; it has also resulted in new research into buffalo hunting and early ranch history.

The U Lazy S Ranch is part of Texas ranching history. In 1901, John Banyon (J. B.) Slaughter bought the Square and Compass Ranch and 5,000 head of cattle, renaming it the U Lazy S. The ranch eventually spread over two counties, Garza and Lynn.[1] Bob Macy, J. B.'s great-grandson and owner of part of the ranch, notes that his family should not be confused with that of Texas John Slaughter, the famed Texas Ranger and cattle rancher. "I can't find anywhere that says that we're related to Texas John," says Macy. The confusion most likely comes from the fact that J. B.'s brother, C.C. (short for Christopher Columbus), established his own ranching empire of more than one million acres, earning him the moniker the Cattle King of Texas.

In 1906, J. B. sold acreage to C.W. Post, who founded the Postum Cereal Company, which eventually became General Foods Corporation. Post bought about 200,000 acres of adjacent land, hoping to create a model city, building homes and other structures, including a hotel, a gin, and a textile plant. By 1909, the town had a school, a bank, a newspaper, and a population of about one thousand. The city changed its name to Post when officials incorporated in 1914, the year Post died.[2]

The original U Lazy S Ranch is now divided between four descendants of John Banyon. Although they renamed their individual portions, the U Lazy S Ranch name still exists because the heirs jointly own mineral rights.

Bob Macy calls his part of the ranch simply Macy Ranch. Although he's not as active as he once was, leaving the day-to-day operations to his son, he still works on the ranch, running the road grader to maintain the ranch roads and flying his helicopter to check on the herds. While Macy loves the ranch, he would not describe himself as a typical rancher. He grew up in Florida, playing competitive tennis. His parents were both professional tennis players. He played the game at Texas Technological College, moving to the ranch when he graduated. "When I was twelve to thirteen years old, I started traveling the country playing tennis," he remembers. "I would stay at the ranch when I had time. While I thought it was playing at being a cowboy, I ended up falling in love with the ranch."

In 2005, Doug Cunningham, who works with the Landmark, introduced Eileen Johnson to Bob Macy and his cousins. "Little by little, we formed a relationship with the Landmark, and I still go out with Doug, and he shows me what he's found and what the crews are working on," Macy says. While Macy has been involved with the Landmark for more than twenty years, his association with the Museum at Texas Tech University began around 1985 when his wife found some tracks in rock and called Sankar Chatterjee, PhD, then the museum's curator of paleontology. While the tracks did not turn out to be of scientific consequence, the episode did lead to Cunningham's taking a job with Chatterjee, working on bones and building replicas for the museum's dinosaur gallery.

Cunningham would eventually take what he calls a curve in the road, moving from paleontology to archaeology. He is a retired Lubbock firefighter, an avid motorcyclist and off-road dirt bike rider, and a generally curious guy. On one of his off-road expeditions in 2004, Cunningham found some objects that he thought might be more interesting to Dr. Johnson than to Dr. Chatterjee, who, as a paleontologist, is focused on ancient fossils, often dating to millions of years before the Quaternary, which is Johnson's focus.

"Doug brought in an artifact and asked what it was and if it was anything important," recalls Johnson, whose office was then down the hall from Chatterjee's." "Turns out it was very, very important." Cunningham found a biface, a stone implement that may have been the first multipurpose tool. People used it to butcher game, scrape hides for clothing, and carve wooden tools. It's not a large tool; Johnson described it as about the size of her hand. The biface Cunninham found was part of a cache of artifacts that had eroded from the side of a draw. "He spotted something different in the landscape, and even though he didn't know what he had, he suspected it might be something," she says. "That first find was all it took for him to look for other things. Now, we have hundreds of locations around the [Macy] ranch."

While Cunningham still spends much of his time scouting for interesting objects and potential excavation sites, primarily around the U Lazy S Ranch, his role with the Landmark has grown. By 2009, Cunningham worked full-time for Dr. Johnson. He is the person who keeps things running during the field season, getting crews back and forth to Lubbock and setting up the camps. He developed the water systems and made other crew-related improvements at the Macy Ranch field camp. When asked how he knew how to do all that, he shrugs and says, "I just figured it out. Now I go to YouTube and learn."

In addition to his site management duties, Cunningham is an integral part of the archaeological team, which values him for his ability to map or relate to a particular place immediately. Johnson

Figure 23. John Moretti, former research aide and senior crew chief (left), Eileen Johnson, and Doug Cunningham discuss excavations at Macy Locality 100 along Spring Creek on the Macy Ranch. (Courtesy of Lubbock Lake Landmark; Stance Hurst, photographer)

is emphatic when she says she has told Cunningham he cannot quit until she retires. "Doug is amazing! I mean, he understands the landscape. He knows the landmarks and remembers what he has seen and where he saw it," Johnson enthuses. "Our hunter-gatherer ancestors had to know how to do this to survive. Although you don't see [this ability] that much in people anymore, Doug can do it. He can visit a location, even one in the middle of nowhere, and return without any issues. He does not get lost."

The Macy Ranch is a particularly rich area yielding numerous significant finds, especially along the Macy Fork of Spring Creek. "Doug found Spring Creek and hundreds of other localities on the Ranch," Hurst says. "I think Doug is responsible for discovering 99 percent of the sites on the ranch. He finds a potentially

Figure 24. Two footbones from a rare American Lion uncovered along Spring Creek Valley on Macy Ranch. These are the first documented remains of the species on the Southern High Plains. (Courtesy of Lubbock Lake Landmark; Jim Gannaway, photographer)

interesting site and asks me to take a look at it. We then determine if the site needs to be excavated."

Over the decades, Landmark crews have made dozens of important finds on or, as Hurst says, within a stone's throw of the area named Macy Locality 100, an area that Johnson calls a premier Quaternary paleontological locality. The locality yielded perhaps the Landmark's most significant find: the bones from an American lion (*Panthera atrox*). Hurst was walking along the Spring Creek valley in 2016 to see if anything new had eroded from the draw's wall and noticed what appeared to be unusual foot bones. He took photos and sent them to Moretti, the Landmark's bone expert at the time. Ultimately, the crew uncovered two bones from a single lion, representing, Moretti says, the first documentation of the American lion on the Southern High Plains.

A few lion fossils have been discovered in places ranging from Canada to Mexico, but researchers usually only find individual bones. La Brea Tar Pits in Los Angeles and Natural Trap Cave in Wyoming are the only places researchers have discovered complete skeletons. At La Brea, thick tar trapped the animals. At Natural Trap Cave, the creatures fell into an eighty-foot-deep pit. In both places, the conditions preserved the lions whole.

Finding American lion bones is rare. Scientific evidence suggests that there were few apex predators during the Quaternary because a large number of prey animals was necessary to support each creature. Johnson knew the lion had an extensive geographic range and suspected the big cats were in the South Plains area; however, the creature had eluded her until Hurst's chance finding. "It had been my dream to find a lion, but it took more than forty years for our crews to discover the bones."

The American lion, which lived from about 200,000 to 13,000 years ago, was the dominant predator in the Americas. Much bigger than today's African lion, the American lion stood about four feet tall at the shoulder and weighed more than five hundred pounds. Despite its size, the lion was fast, estimated to run about thirty miles per hour. Among carnivores in North America, few, including the prehistoric grizzly bear (an omnivore), and the extinct short-faced bear (a scavenger rather than a hunter), were larger than the lion.[3]

The American lion appears to have had a pride-based social structure, indicating communal hunting, which means that almost any animal could have been on the menu. "Based on isotopic evidence of what an animal was eating, American lions primarily ate ancient bison, camel, and horses, but ground sloth and mammoth also were part of their diet," Johnson explains. "The American lion was capable of killing prey larger than themselves. Their stealth and size allowed them to take down large prey and defend those kills against other large carnivores such as the Smilodon or saber-tooth cat and the short-faced bear."

Glen Evans and Grayson Meade, who excavated at Lubbock Lake in the 1940s and 1950s, found the first trace of the American lion in Texas in the Panhandle in 1945. Moretti says the two published a scientific paper detailing their many discoveries in the area, including a mention of the lower jawbone of an American lion that landowners in the area had shown them. Evans and Meade later found additional evidence of the lion inside a rock shelter called Kincaid Shelter near Uvalde, Texas, on the Texas-Mexico border. They excavated the site in 1948, finding, among several other fossils, an upper canine from the American lion. While Evans and Meade never published their discovery, Johnson and Moretti recognized how closely it related to their own discovery, and they decided to undertake and publish an analysis.[4] Johnson explains, "We could document their work for posterity through ours."

The field crew continues to hunt for more of the big cat's bones at the Spring Creek location. They have begun excavating at the top of the deposits, about 10.8 feet above where the bones eroded out. "Eventually, the field crew will reach the bone level and determine what may still be preserved from the lion's skeleton," she explains. Visitors can view a replica of the lion skeleton at the Landmark's Interpretive Center. A second model is in the Ice Age of the Southern Plains Gallery at the Museum at Texas Tech University.

Figure 25. Doug Cunningham assembled the models, with his cat offering inspiration. (Courtesy of Lubbock Lake Landmark; Jim Gannaway, photographer)

In addition to his fieldwork, assembling the lion replica fell to Cunningham. He has also built replicas of a Smilodon, a saber-toothed cat, and several creatures in the museum's A Changing World Gallery. "He's a master at that sort of thing," Johnson says. "He can really make the bones come alive."

Landmark crews often use plaster casts when excavating fragile bones, but Moretti's team was tested in 2013 when they recovered a complete giant tortoise shell. The shell spanned about three feet across and dated to the late Pleistocene. The massive creatures would have lived on land, similar to modern-day Galapagos Island tortoises.

The crew discovered the shell in a modern drainage channel up the slope from the main Macy Locality 100 excavation area. Using shovels, the crew dug down to the sediment housing the shell. Over several days, they used wooden picks and small brushes,

Figure 26. In 2013, the Landmark crew recovered an intact giant tortoise shell about three feet across. Plaster casts, like this one, are often made to transport fragile bones and artifacts. (Courtesy of Lubbock Lake Landmark; Jim Gannaway, photographer)

expecting to discover pieces of the shell, but instead they found the entire thing. Along with the shell, the crew found plates of bony armor embedded in the skin over the limbs, indicating that the tortoise's skeleton might remain inside the shell. The armor, which is not part of the shell, forms a protective layer over the legs. The spikes on the armor resemble upside-down Hershey's kisses. "The intact condition of the tortoise made it a significant discovery from any time in antiquity," says Moretti.

The crew created a huge three-and-a-half-foot plaster jacket around the shell and surrounding sediment, weighing eight hundred to one thousand pounds combined. The team slid a steel plate under the jacket so that it could be loaded onto a small trailer pulled by an ATV. It was then transferred to a truck and taken to the museum for future laboratory work.

One Macy 100 excavation yielded much more than the Landmark crew expected. In 2012, Cunningham spotted an ancient bison horn on one of his explorations. In addition to the bison horn, the excavation uncovered the remains of a western camel (*Camelops hesternus*). Camels originated in the New World and routinely grazed the Southern Plains. It was a significant discovery yielding a skull, an almost complete neck, and two leg bones. "The bones are important finds that could contribute to understanding ancient animals and the regional environment," Moretti says.

Although crews have discovered many fossils much older than those dating to the Quaternary Period, which is their focus, the oldest fossil award goes to a rhinoceros mandible, found buried in Ogallala Formation deposits. Cunningham saw the bone in 2009 on Macy's property. Researchers, who spent a day excavating the bone, estimate it is more than eight million years old, the oldest object Landmark crews have ever unearthed. It is one of the many finds representing the unusual animals that people today do not expect to have lived near Lubbock. While much older than the Landmark's research program timeframe, Moretti believed the find was too important to pass up.

COWHEAD MESA

One of the Landmark crew's more unusual endeavors involved preserving the petroglyphs at Cowhead Mesa on Macy Ranch. On closer examination, what appear as scratches on the sandstone walls are actually representations of centuries-old Native American and Anglo-American cultures. People have known about Cowhead Mesa for decades. Macy remembers riding with the cowboys to the area as a teen. Landmark crews have worked at the site since 2005. In 2008, Hurst and Johnson revisited the mesa as part of a more extensive study on how hunter-gatherer

Figure 27. Cowhead Mesa on the Macy Ranch presented an unusual project for the Landmark: preserving the fading petroglyphs, or rock art, on the sandstone wall dating from the 1600s. (Courtesy of Lubbock Lake Landmark)

groups used the landscape of the Southern High Plains. On top of the mesa, they surveyed the area and mapped and collected artifacts. They recorded four hearth features and one rock cairn.

Johnson excavated two hearths to see how they were constructed and to gather material for radiocarbon dating. In a 2017 article in the *Plains Anthropologist*, Hurst and Johnson wrote that amateur archaeologists had previously noted the cultural significance and prominence of Cowhead Mesa.[5] "Rock art is rare in the region, so this was an important opportunity for us," Hurst explains. "The archaeology on top has allowed us to examine a hunter-gatherer campsite associated with a unique landscape feature."

While most cultures name important sites, no such native name has been found for Cowhead Mesa. Hurst and Johnson note in their article that "local folklore suggests that an early Anglo rancher may have named Cowhead Mesa for a bison petroglyph attributed to the historic Comanches."

One large and nine smaller panels comprise the petroglyphs, dating between 1600 and 1800, roughly covering the Apache and Comanche occupations of the area. The Native American art depicts, among many other things, two multistory missions, with one appearing to be on fire; geometric shapes; animals; and human figures, some wearing headdresses and holding weapons. The 1881 cattle brand of one of the earliest ranches in the area, the Square and Compass, the ranch John Banyon Slaughter bought, is also depicted. Multiple layers of rock art are visible, one on top of another. Unfortunately, modern graffiti on the walls also obscures some of the art underneath.[6]

Hurst and Johnson were particularly concerned about the deterioration of the rock art. With no way to stop the erosional process, they called on Elizabeth Louden to help preserve the petroglyphs. Previous researchers had tried to document the art, but early drawings, rubbings, and photographs were of poor quality. Laser scans and photographic techniques can reproduce the petroglyphs on the nine smaller panels that comprise the rock art area; however, none of these techniques would be effective on a single large panel.[7]

Figure 28. To better preserve the petroglyphs, the Landmark joined forces with an architecture professor to digitally photograph and laser scan each of the nine small and one large rock art panels. (Courtesy of Lubbock Lake Landmark)

Louden, then a professor at the Texas Tech College of Architecture, had used long-range scanners to develop the first accurate measurements of the exterior of the Statue of Liberty in 2006. The National Park Service had engineering drawings of the interior but not the exterior. She also worked on laser scans of the Spring House Pueblo at Mesa Verde National Park in New Mexico, George Washington's tent at Valley Forge National Park, the Roman Forum, and Luxor, Egypt.[8]

With Louden on board, the Landmark team set out to create a model to preserve the entire context of the Cowhead Mesa panel and document each of the art elements to determine how these petroglyphs compare to others in the region. They captured a series of close-up, high-resolution digital photographs of the large panel and each of the smaller ones.

While Macy is interested in the ancient animals, he is especially fascinated by the history of the region's early ranchers and buffalo hunters who operated there. "Even though we're talking only as far back as the 1860s or '70s, it's much more interesting to me," he explains. Like the ancient animal discoveries, the U Lazy S Ranch has proven a rich source of historical archaeology, opening up remarkable new research opportunities for the Landmark. Hurst notes that he started his career focused on Paleoindian research, but working on the ranch has changed his perspective. "I wasn't interested in historical archaeology at all," he says, smiling. "But working with Dr. Johnson has broadened my horizons."

The sites on the ranch are pristine, likely because they are so remote. Hurst explains that other sites often have modern trash and debris covering up historical artifacts. The fact that the U Lazy S Ranch sites are off the beaten path, he thinks, has resulted in their unspoiled condition. "We've made significant discoveries related to the first cattle ranchers and the buffalo hunters who worked in the region," he says.

One of the first early ranch history discoveries is what Hurst and Cunningham believe is a cowboy roundup camp dating back to the mid-1880s. They uncovered several .45 Long Colt cartridge cases using a metal detector. Cunningham explains that the brass case holds the bullet, gunpowder, and a primer, which fits into the gun's chamber. After firing, a cartridge casing is ejected from the gun. There were so many cartridge cases that Hurst theorizes the cowboys were engaging in target practice or possibly participating in a shooting competition.[9] They also located what appeared to be a section of a chuck wagon, including a side coffee grinder designed to mount onto a wagon.

Macy played a role in finding the first of two dugouts on the ranch. He was flying his helicopter over the area and noticed a chimney and the remnants of a dugout. The dugout is significant because it reveals both buffalo-hunter and early open-range ranching history. Dugouts were the first shelters Anglo-Americans built when they settled on the Plains; how the dugout was made

Figure 29. Eileen Johnson and former field assistant Scott Garrold stand in part of a buffalo hunter dugout on the Macy Ranch. Former student Olivia Rovelli is in the background. (Courtesy of Lubbock Lake Landmark; Stance Hurst, photographer)

depended on the availability of resources and landscape. Settlers often built the structures into the side of a hill or ravine where possible. If the landscape was flat, they dug a deep pit. Builders often constructed front entrances from sod or logs.[10]

Using a metal detector and excavating in and around the dugout, Cunningham and Hurst found numerous objects at the location, including cooking utensils, metal food containers, and pieces of glass bottles, including an intact medicine bottle. Cartridge cases located at the dugout indicate the presence of several different rifles or revolvers.

BUFFALO HUNTERS

While early ranches and buffalo hunting might seem far removed from the Landmark's usual focus, Johnson emphasizes the importance, particularly of the buffalo hunters, in the region's archaeological record. "Buffalo hunters are a part of the record largely missing at the Landmark," she explains. "We have a buffalo skinning knife from the Landmark, but that is insufficient. We have nothing concrete to say the hunters were at the Landmark, but they must have come through here as they moved south."

The modern American bison is commonly referred to as the buffalo (scientific name *Bison bison*). "Buffalo" is the name French fur trappers gave to the creatures in the early 1600s, from the French word *boeuf*.[11] The hunters played a vital role in paving the way for early ranchers and settlers moving to the Plains from the east.

Estimates suggest that around sixty million bison roamed the United States in 1870; twenty years later, that number dropped to around five hundred. As the railroad was completed through the western part of the United States, settlers began moving into the Great Plains in growing numbers. Trains carried millions of hides east. The hunts began when the bison's fur became popular in the East for coats and robes. Factory owners also discovered that the hides could make stronger machine belts. Most critically, the hunts met the needs of the US government, as some officials urged buffalo hunters to wipe out the animals to force the Comanches and other tribes onto reservations in Oklahoma.[12] Buffalo hunting began as a primarily economic endeavor and peaked in the late 1880s, with the near-total extinction of the animals.

Johnson notes that while there is abundant historical research and documents that focus on telling the story of the buffalo hunters, she was surprised at the relative scarcity of scholarly anthropological research. Only three buffalo-hunter dugouts have been excavated in Texas. The Landmark has two dugouts under excavation and one that has been surveyed, but as of 2024, excavation has not yet begun.

There is a wealth of information on buffalo hunters, some factual and some embellished by the hunter or those who have told the stories over the years. Numerous articles populate both academic and popular literature. Many hunters kept diaries and gave interviews to journalists and historians. Some of the hunters even wrote books. "We are drawing from those stories. One of the hunters gave a detailed account of how the military tracked the Comanche over several months. That operation eventually ended in the Red River War," Johnson says. "However, what is important from our perspective is that he gives distances and names of various landmarks. Important information that can lead us to archaeological evidence of what happened."

The Red River War (1874–1875) was a conflict in which US Army troops launched numerous campaigns to remove the Native American tribes from the Southern Plains to reservations in present-day Oklahoma. Western expansion and the massive killing of the buffalo, which the tribes depended upon for survival, sparked raids on settlers and buffalo hunters. While the tribes asked the government for help, that same government was actively encouraging western expansion.[13] James L. Haley writes in *The Buffalo War: The History of the Red River Indian Uprising of 1874*, "It [the war] saw the extermination of the bison from the South Plains, and it safeguarded the opening of the territory from central Kansas to central Texas to white settlement."[14] The war effectively ended the more than 13,000-year presence of Native Americans on the South Plains.

The Landmark's archival research turned up correspondence by buffalo hunter J. W. Woody to noted historian J. Evetts Haley. Woody writes that he, his brother, and two other hunters built a dugout with a rock chimney, their headquarters near Post in Garza County. "We were camped at an old Indian town," he writes. "There were 172 teepees in it, made of poles and covered with grass."[15] In a February 11, 1928, letter to Haley, Woody notes that his party killed 3,200 buffalo during the winter of 1876 and spring of 1877.[16]

Much like one sees with ballistics tests on CSI television shows, the Landmark researchers found unique marks on cartridge cases, which are ejected from the gun when it is fired. Hurst and Cunningham believe that several .50 caliber cartridge cases fired from a Sharps rifle were not fired on-site but elsewhere, and the cases were brought back to the dugout for reloading. The Sharps was the rifle of choice for buffalo hunters.

The Woody brothers eventually abandoned the site, and Hurst says J. H. Lindsey, the first open-range cattle rancher in the region, used the dugout as his headquarters from 1879 to 1881. The Landmark team found Lindsey's brand at the dugout, indicating that he and, most likely, his family occupied it. The brand was confirmed as Lindsey's through a search of cattle brand records at the Garza County Courthouse.

Hurst and Cunningham found a second dugout near Salt Creek in Garza County, which they believe belonged to buffalo hunter Marshall Sewell. A severe food shortage at the reservation in Fort Sill, Oklahoma, had prompted Comanche war chief Black Horse to request and receive a permit from the reservation agent to hunt bison in Texas to bring back to the reservation. However, as the Texas State Historical Association website notes, the decimation of the bison herds angered Black Horse, and he planned to attack every buffalo hunter he found. Marshall Sewell ended up in Black Horse's sights.[17]

A captive of the Comanches, Herman Lehmann, recounts that Black Horse's party found a lone hunter shooting buffalo. In his book, *Nine Years Among the Indians, 1870–1879*, Lehmann writes that the hunting party "surrounded the hunter, killed, and scalped him, and then took after the two who were coming from the camp." The two buffalo skinners who saw Sewell's killing ran from the Comanches. Unable to find the two skinners, Lehmann writes, the hunting party returned to the dugout and destroyed everything they could find.[18]

The skinners whom Sewell had hired, Wild Skillet and Moccasin Jim,[19] went to Rath City, a short-lived frontier town that served as the main supply point for the region. By early March 1877, about forty men rode out of Rath City and tracked the Comanches. There are conflicting accounts of the events that follow. Landmark researchers have sought archaeological evidence of what is now known as the Battle of Yellowhouse Canyon but have found none to verify any single report.

One account in the *Handbook of Texas* on the Texas State Historical Association website states that the group picked up the Comanches' trail at Sewell's camp and trailed them to Yellowhouse Canyon. Although the Comanches and the buffalo hunters skirmished, neither side could claim victory. Word of the battle reached the US Cavalry, who eventually caught up with the Comanches, forcing a surrender and returning them to the reservation.[20]

There are also various theories as to Sewell's murder, including that it occurred in what is now Mackenzie State Park in Lubbock. However, Hurst and Cunningham believe that the site they are working on is where Sewell met his death if, as Hurst notes, the written accounts they have drawn from are all true. He and Cunningham have examined many stories about the same event, some with differing details. "I guess there are four or five accounts, and they all conflict somehow," he says. "That's where the archaeology comes in. The historical accounts are an important aspect, but the archaeology tells a story that is not in the historical record." Their archaeological research indicates that Sewell was killed at the mound where he shot the bison.

Hurst says they have not found Sewell's remains, but he is confident they have the location from which Sewell was shooting. Based on locales and other details from their historical research, including a firsthand account from J. Wright Mooar, one of the most famous buffalo hunters, Cunningham went in search of the location that Mooar pinpointed as an outcropping known as the Devil's Breakfast Table on the U Lazy S Ranch. "When I read the account, the hair on the back of my neck stood up because I knew where that place was," Cunningham says.

Cunningham took a metal detector to the spot Mooar described as where Sewell was shooting buffalo and found a .45 Long Colt cartridge case dating to around the late 1870s and several square-cut nails. In a second investigation at Sewell's dugout, the researchers found .50 caliber rifle cartridge cases and two Comanche metal arrowheads.

As Cunningham thought about the mound, he realized there might be bullets near where the bison herd was grazing. Cunningham's research told him that about twenty bison were lying near the site where he believes Sewell was killed. He knew if the animals were there and had not been skinned, he could find bullets. A bullet shot from a Sharps rifle could reach anywhere from 400 to 800 yards with accuracy. In his research, Cunningham found one description that noted the sweet spot from which to fire at bison was about two hundred yards. Any closer and the sound would scare the herd. "That's where I started looking," he explained. Borrowing a friend's rangefinder, he looked 200 yards from the mound. "And bam! I found four bullets." Cunningham eventually discovered ten bullets in total. The fieldwork on this project has concluded, and publication of an academic article on the project is pending.

Cunningham's enthusiasm for and understanding of the territory make him an invaluable member of the Landmark team. Cunningham is self-effacing, describing himself as a Southern California guy who ended up in West Texas after leaving the Air Force. Although he portrays himself as someone without much formal training, Hurst views him as an integral part of the team. "I consider him a researcher," Hurst insists. "He may not be a professionally trained archaeologist, but he has a good view of things and a different perspective, which I appreciate. He heavily influences so many things that we do."

While other sites have not produced the same breadth of discoveries as those found at the U Lazy S, excavations across the surrounding region have yielded their own significant finds. Together, these regional sites help place the Landmark's discoveries in a broader context, providing a more complete picture of the animals and humans who once lived on the South Plains.

CHAPTER 6

A REGIONAL CONTEXT

Eileen Johnson began exploring the broader South Plains region almost from the start of her tenure, and each site or locality reinforced her decision to explore the entire Quaternary Period. "I have learned not to limit my view," she explains. "While we began looking at early human habitation, we have since found much older animal fossils and more recent artifacts denoting the presence of American Indians in the area, buffalo hunters, and the beginnings of the City of Lubbock."

Excavations in the region surrounding Lubbock help place the Landmark's discoveries into a broader context, providing a more complete picture of the animals and humans who once lived on the South Plains. Yet Johnson's work extends well beyond the Landmark and the immediate region. Her collaborations reach across the Americas.

Vance Holliday, one of Johnson's longtime colleagues, believes her reputation in the archaeological world began with her 1975 doctoral dissertation from Texas Tech University and two archaeology symposia she organized, one in 1974 and the second in

1977. "[The] symposia sessions were well attended by Paleoindian archaeologists, geologists, and paleobiologists putting Eileen onto the national and international scene," Holliday wrote in a volume of *Quaternary International* honoring Johnson.[1]

By the 1980s, Johnson was receiving invitations to collaborate on projects exploring the grassland environment, butchering sites, and bone tools in the Pampas of Argentina. That research began in 1991. Johnson brought one of her Argentinian colleagues to Texas Tech in 1996 to work on the field crew and, later, to complete a master's degree.

At about the same time, Johnson received a grant from the National Geographic Society for renewed excavations at San Josecito Cave in Mexico. The work was research for a doctoral student working with Johnson. He returned to Mexico to head the Laboratorio de Arqueozoología at the Instituto Nacional de Antropología e Historia. The research eventually led to the establishment of the Mexican Quaternary Mammal Database.

In 1994, archaeologist David Overstreet of Marquette University asked Johnson to participate in a Wisconsin project analyzing a mammoth's bones for chemical or physical changes after the animal died and how humans had modified them. The bones date to about 14,000 years old, making the find pre-Clovis.

Johnson says that her experiences working across the Americas have helped her to better understand the Landmark's records and fit them into a broader picture. That mindset has carried through to various regional sites and localities around the Landmark. Her more than half a century at the Landmark has refined her perspective. "Over the length of time I've invested here, I see things differently," she explains. "Things we thought we knew at the beginning, we now know weren't what we thought. However, we can go back and look at an excavation we conducted then or open a new site adjacent to it to understand better what was going on and what it means."

Johnson's international research has taught her that some things will overlap despite thousands of miles between sites. "I was amazed at how much the Argentinian pampas reminded me of the West Texas prairie," she says. "We found a ginger beer bottle in Argentina similar to one discovered at the Landmark."

The Landmark's regional research began almost immediately upon Johnson's arrival, with her participating in a reexamination of the Lake Theo site near Quitaque, Texas, now in Caprock Canyons State Park. The site, first discovered in 1972, reveals a bison bone bed that researchers believe was a Folsom bison kill and processing site owing to recovered Folsom points. Johnson designed her 1977 research to determine the extent of the site, examine the stratigraphy and buried soils, and recover microvertebrate remains.

In the 1990s, Johnson worked with graduate students on an investigation at the Ryan site near Shallowater, Texas, northwest of Lubbock. The site holds Plainview-age lithics. She has also examined bones from key Paleoindian sites, including those at Plainview, Texas, and Clovis, New Mexico.

Over the last half-century, many locations have produced new information for Landmark crews. As with everything, times change; older sites close, and new ones open. Excavations have occurred from as far west as Eastern New Mexico to as far south as Abilene, Post, Snyder, and the Permian Basin.

ADAIR-STEADMAN

The Adair-Steadman site near Roby, Texas, south of Lubbock, is a notable example of how early humans adapted to changing climates and landscapes. The site is one of the first and largest spearpoint production areas, where Folsom points were made for hunters to carry with them rather than making points as they traveled.[2]

The location is a primary Folsom hunter-gatherer site on a terrace north of the Clear Fork of the Brazos River on the Texas

Rolling Plains. It is named for the landowner Link Adair and Foy Steadman, a local postal delivery person.

Estimates suggest that it would take humans living in the early Holocene about seventy-six hours to walk from Adair-Steadman to Lubbock Lake, which scientists know was a favored bison hunting ground. Today, the same 119-mile trip takes less than two hours by car. Since Adair-Steadman and Lubbock Lake are located in the Brazos River Basin, the Landmark crew theorizes that some of the Folsom points uncovered at Lubbock Lake could have come from people moving along the river valley from Adair-Steadman.

The first excavations, by the State Archaeologist of Texas Curtis Tunnell, ran from 1969 to 1974. His crew uncovered more than 45,000 stone tools and flakes. Flakes are pieces chipped away from the stone during the tool-making process. Scholars credit Tunnell as the first scientist to figure out how Folsom projectile points were made.[3]

The process of making the points and tools is known as flintknapping. Flintknappers used a soft hammer, usually an antler billet, to hit the stone to fracture a rock at an angle. Then, a flaker, usually an antler tine, created narrow grooves in the rock's surface. Folsom hunters wanted lightweight tools and weapons that could perform several functions. The points thus fashioned were thin, light, extremely sharp, and considered among the most reliable big-game hunting points ever made.[4]

The Landmark team has worked at the Adair-Steadman location since 1999. Their initial efforts included reexamining objects from the original investigation in search of new information on Folsom flintknapping. Most recently, they have returned annually for one-day pedestrian surveys, walking the site and occasionally finding new objects. In 2017, the team used a drone to map the site and create a 3D model so that researchers could better monitor future changes to the area's surface.[5]

Figure 30. In 2017, Stance Hurst deployed a drone to map Adair-Steadman. A 3D model was created so that researchers could better monitor future changes to the area's surface. (Courtesy of Lubbock Lake Landmark; Jim Gannaway, photographer)

The Landmark's staff developed an online exhibit based on a 2017 short-term physical exhibit. The online presentation, on the Landmark's website, describes the flintknapping process. The exhibit also traces how Folsom people traveled on foot across the Southern High Plains, following rivers and carrying their tools and weapons, and features videos and still photographs describing the research at Adair-Steadman, toolmaking, and 3D processes used by the Landmark staff.[6]

YATER RANCH

Yater Ranch offers a combination of prehistoric and historic objects. Lyn Yater, who owns the ranch near Stanton, Texas, in the Permian Basin, became involved with the Landmark because of her curiosity and love of history and archaeology. Her father was a geologist, and she had always liked rocks and being outside. Her late husband knew archaeological objects were on the ranch but, she says, he did not care. "I did care, and I thought I needed to learn more," she says. "I tried to take an archaeology class at Midland College with Eunice Barkes, but not enough students wanted to take the class that semester, so the college canceled it. I was lucky that she said if I were willing to help her with the collection at the Museum of the Southwest in Midland, she would teach me as we went."

Barkes was interested in Randolph Marcy, a US Army officer and Western explorer. In 1852, Marcy led an expedition across the Great Plains, looking for the headwaters of the Red River. Later, government officials called Marcy to Washington, DC, to write a guidebook for people heading west, which was titled *The Prairie Traveler: A Hand-Book of Overland Expeditions* and published in 1859.[7]

When Barkes realized Marcy had come through part of what would become the Yater Ranch, she wanted to see the area. After looking around, she called on her friend, Eileen Johnson, at the Landmark, asking her to look at the location. Johnson and some of her crew made an informal visit to the ranch in late 2008. "It was December, and I think the temperature was about 23 degrees, but I didn't care," Yater says. "I was as happy as I could be. There is a world of stuff to uncover out here."

Figure 31. Excavations on Yater Ranch have produced evidence of human habitation from the Archaic to the Historic Period. A large modern bison and horse bone bed has been uncovered in Mustang Draw. It is the second known combined bison and horse kill site. The first is at the Landmark. (Courtesy of Lubbock Lake Landmark; Jim Gannaway, photographer)

Johnson was also happy with the site, and a crew began work in 2009. Yater is fascinated by the archaeological work and is often on-site to see what the crew finds. "I like to know the whole story about anything we find," she says. For Yater, learning everything she can about archaeology and the ranch she loves is exciting. She has been on the ranch for more than thirty years, but her husband's family has owned it for more than a century; his grandfather bought the land in 1923. "I was always interested in ranching, and I hung around with cowboys, so moving to the ranch when I got married was a dream," she says. "I think knowing your past is important. And I want to know about other people's pasts, too. Why did they do this or that? I was lucky to get to live on a ranch."

Yater's knowledge of ranches and their history also leads her to watch Westerns on television a little differently than most folks. She collects saddles and spurs and knows what she is talking about. She tells the story of a friend asking her what kind of spurs Festus, a character on *Gunsmoke*, wore. The question sent her off on the type of research project she loves. "I had to find a good picture of the spurs," she explains. "I think I have narrowed it down to Crockett spurs. They were common back then, so that is probably what they were." Whereas *Gunsmoke* scored well in accuracy on the spurs, other shows missed the mark. "I've noticed shows where the cattle are wearing modern-day ear tags," she says, shaking her head. "I don't pay as much attention to the story as I do the saddles and spurs and clothes to see if they're right."

There are two sites on the Yater Ranch that have yielded significant animal bones and evidence of human habitation ranging from the late Archaic Period to Comanche occupation during the Historic Period. One site overlooks Mustang Draw and is referred to as the Arena. The second site is located within Mustang Draw at Mustang Pond.

Lyn Yater wasn't the only one watching the work at the Mustang Pond site during the 2013 season. The ranch horses grazing in the nearby pastures also kept a close eye on the activity. An exceptionally inquisitive horse, Whiskey, took his oversight duties seriously.

Figure 32. Whiskey the horse checks the work of one of the field crew at Mustang Draw. The Landmark crew calls the site at Mustang Draw Whiskey Flats in his honor. (Courtesy of Lubbock Lake Landmark; Stance Hurst, photographer)

"One day, we had just excavated several horse bones and had moved them temporarily from the excavation site," says Hurst. "We turned away for just a minute, and when we looked back, Whiskey was licking them." What Whiskey could not know was that the bones belonged to his ancestors. Hurst theorizes the bones may have smelled different since they had just been dug up. Whiskey continued to oversee the dig for several years.

Archaeologists assign names and numbers to excavation sites that follow a trinomial (three-part) system, which the Smithsonian Institution developed in the 1930s and 1940s. The Smithsonian no longer maintains the system; each state assigns its own trinomial. Locations not registered with the state are usually classified as paleontological localities rather than archaeological sites.[8]

Archaeological sites contain human-made artifacts and provide insight into the history of human cultures. Paleontological localities consist of plant and animal fossils that allow scientists to study the history of life on Earth. The Landmark researchers give sites and localities a name, usually the name or initial of a specific ranch, and a number.

The trinomial designation for archaeological sites consists of a state designation (in Texas, it is 41), a county designation (MD for Midland County), and a consecutive number; for example, a site could be registered as 41MD50. The number 50 would denote the Mustang Draw location.[9] The Landmark crew named Whiskey Flats, located at Mustang Pond, after the inquisitive horse.

While the bones in Mustang Draw are those of modern horses, there were once dozens of species of horses (family: Equidae; genus: *Equus*) roaming the Earth. Horses originated in North America about 55 million years ago. Before they became extinct in the New World, many migrated to Europe and Asia via land bridges. The Spanish and other Europeans later brought horses back to the Americas in their domesticated, modern form.[10]

As noted, Mustang Pond is the region's second-known combined bison and horse kill site. The first is at the Landmark. "We're finding things in the region that add to the record we have established at the Landmark," Johnson says. She emphasizes that the record at the Landmark dates back 2.6 million years. Although people were not present that far back, centuries of soil deposits preserved the record. "This is what makes the Landmark so important," she adds.

Mustang Draw has yielded much more than animal bones: The top of the draw has produced dozens of hearth features and other artifacts that give researchers a glimpse into the lives of humans who lived in the area, dating back approximately 4,500 years ago. It has, so far, produced the earliest materials that date to the Late Archaic Period. Hearths were firepits that could be left as dirt or lined with a wide range of materials such as stone, which people used for cooking and warmth. Among the significant discoveries are about 4,000 artifacts and bones dating back to at least the Comanches, who dominated the Southern Great Plains from the mid-1700s to the late 1800s. Excavations along the rim of Mustang Draw have revealed an extensive Comanche village, including living areas that Hurst compares to modern living rooms. These areas include hearthstones, charcoal, and other camping artifacts people left behind. Johnson notes that in the records of his explorations, Marcy remarks on a large Comanche encampment on the rim of Mustang Draw, overlooking Mustang Pond.

SLATON RESEARCH STATION

A newly opened location already adds to Hurst and Cunningham's buffalo hunter research specifically, and it has the potential to contribute much more to the Quaternary generally. Initial work began during the 2024 field season at a dugout on a ranch near Slaton, Texas, east of Lubbock at the mouth of Yellowhouse Canyon. The landowners requested that the Landmark not name the ranch.

The landowners brought their ranch to the Landmark's attention. "While visiting the Landmark, the family saw the buffalo hunter exhibition, which features information about the dugouts we've found near Post," says Johnson. "The family talked to our staff on duty at the Interpretive Center that day and said, 'We may have one of those on our land.' That led to discussions with Stance, and we signed an agreement in 2024."

There was indeed a dugout on the property, which Hurst hopes will add to the history of buffalo hunters in this area. He notes that the crew only spent three to four days surveying the dugout in 2024, but it is already yielding results. Besides finding shell casings and other metal artifacts, they discovered something they hadn't encountered at the Post sites: lead slag. The slag indicates that the buffalo hunters were melting down used cartridges and making new bullets.

This dugout is much larger than those found near Post, and Hurst says there are indications that multiple groups used it over the years. As of 2026, crews are still exploring the dugout and the area immediately around it. They hope to find more artifacts and possibly some indication of who lived in the dugout. "It's going to be a big, long-term investigation based on the size of the dugout and the fact that it is pretty well buried."

After the 2024 field season ended, Cunningham spent time on the ranch near Slaton doing reconnaissance work, looking for bones or artifacts on the surface that could lead to other potential excavations. Johnson believes there is potential for various exciting discoveries over the next several years. The owners have primarily used the land for grazing cattle, and except for some oil wells and gravel mining, it has largely been undisturbed.

The landscape is also very different from other locations. For example, in Post, the land is crisscrossed with draws, which means there is more erosion, making it easier to see bones protruding from the sides of the draws. The Slaton location sits on the edge of the Caprock, with a gentle slope and a series of terraces rather than a steep drop-off to the canyon floor.

The landscape speaks to Johnson's love of the prairie landscape. "It is a beautiful vista, just a magnificent view," she enthuses. "Yellowhouse Creek, which runs near the dugout, has water. Another creek flows closer to the dugout and is spring-fed."

The water and gentle slope make for what Hurst theorizes could have been a highway for bison herds moving on and off the Caprock. The area could also have been a pathway for early humans and ancient animals. Hurst and Johnson are excited that the location has the potential to yield amazing finds such as those at the Landmark or on the Post site. "The U Lazy S Ranch is on the South Fork of the Brazos River," Hurst explains. "The Slaton location is on the North Fork. If you follow the water from Slaton, you'll run into the Landmark."

ROLAND SPRINGS

The Roland Springs locality and the Turtle Creek tributary of the ancient Clear Fork of the Brazos River, which runs through the area, have yielded fossils of more than fifty identifiable species of fish, birds, reptiles, amphibians, and mammals. Roland Springs appears to be earlier Pleistocene (2.6 million to 11,000 years ago) in age. The extinct pygmy Aztlan rabbit (*Aztlanolagus agilis*) is present in the recovered fauna and is known only from the Pleistocene. The locality, near Snyder, is significant not only because it has yielded fossils but also, as Moretti explains, because it reveals the known habitat preferences and dietary requirements of those animals, which provide insight into the diversity, landscape, and environments of ancient West Texas.

The creatures in the area included rabbits now extinct with no living relatives and a progressive hare related to modern-day jackrabbits. One of the more interesting mammals occupying the area was the ancient miniature three-toed gazelle horse. Excavations also revealed the jawbone of a large cat that Moretti believes is related to the puma and a cheetah-like cat, but not the American lion.

During the 2012 season, Moretti's crew uncovered an extinct proboscidean tusk at the Roland Springs location. Proboscideans are a group of animals with trunks, such as elephants, mammoths, and mastodons. The ancient *Zygolophodon*, from the Mammutidae family that includes mastodons, is thought by scientists to have arrived in North America from Eurasia about 17.5 to 16.5 million years ago. The American mastodon (*Mammut americanum*) is endemic to North America and is believed to have evolved from *Zygolophodon*.

Moretti writes in the winter 2013 edition of the Landmark's newsletter that the find is the first definitive evidence of an extinct proboscidean in the known fauna of the Roland Springs locality. The tusk indicates that there were trees, brush, and grasslands in the area. The giant elephant-like creatures usually ate tree leaves and brush, suggesting that while not overly dense, the wooded area extended much farther than researchers previously thought.[11] Finding remains of extinct turkeys, giant tortoises, zebra-like horses, and small three-toed horses helps researchers better understand the local environment based on habitat preferences and dietary requirements.

SAN JON

San Jon sits in an ancient playa basin in Northeastern New Mexico. It is one of only a few stratified playa sites in the Southern High Plains area that show in situ records where different peoples occupied the area over multiple time periods.

The area reveals occupation throughout the late Quaternary, from Paleoindians to Historic-era Anglo settlers. A lanceolate point, which looks like the tip of a spear, indicates a camp area dating from about 9,800 years ago during the Early Archaic. Johnson explains that while Landmark crews have recovered evidence from the Archaic Period, Early Archaic Period sites are rare, with only two excavated for the entire Southern High Plains. San Jon is one, and the other is at the Lubbock Lake Landmark. Landmark researchers spent several years reexamining existing materials and conducting new excavations at the San Jon locality. Landmark crews began work in 1999 and completed it in 2013.

TAHOKA LAKE

Tahoka Lake has revealed a wealth of information from more modern times. Tahoka Lake is about thirty miles southeast of Lubbock, just outside the small town of Wilson. It is a salina, one of about forty saltwater lakes on the Llano Estacado. Mammoth Creek flows into the north end of the lake and, with other springs in the area, provides a mix of freshwater into the salina.

The lake is home to archaeological evidence and early ranch history. The ranch became known as the Tahoka Lake Ranch in the mid-to-late 1890s and at the time was owned by Christopher Columbus (C. C.) Slaughter. From 1897 through 1907, Slaughter's foreman and manager, Jack Alley, had his headquarters at the lake. Mrs. Clyde May has lived on the land since January 1996,

Mrs. May has allowed the Landmark crews to work around the lake since 2001. Most of the work is centered on a terrace, a flat piece of land that provides a view of the lake, and along Mammoth Creek. The lake is partially dry during droughts. "It looks like a snowfield out there," she says. "But it is the salt crust. I have broken little pieces out of the crust and tasted them. It tastes just like the salt you buy at the store."

Mrs. May is rightly proud of the various roles her land and the lake have played over the centuries. "The history is really rich because of the people who were here," she says. "You can come out here and see how the different cultures overlapped."

In addition to finding 30,000- to 40,000-year-old animal fossils, mainly the remains of horses now extinct, crews have discovered the remains of a rock corral for sheep, making it one of a handful of *pastores* sites this far south. Beginning in the 1870s and continuing through the mid-1880s, the *pastores* (Spanish for sheepherders) migrated from New Mexico, settling along the Canadian River in the Texas Panhandle.[12] While most *pastores* sites are located in the Canadian River Valley, the sheepherders also moved their flocks down the eastern base of the Llano Estacado into the canyons as far south as Yellowhouse Canyon.

The plentiful grasslands and water that remained after the removal of Native American tribes from the Panhandle attracted sheep owners to relocate their flocks from the New Mexico range, which could no longer support the ever-growing numbers. J. Kent Hicks and Eileen Johnson, in a 2014 article for the Society for Historical Archaeology, note that in 1932, Curry Holden described

Figure 33. A Landmark crew examines Tahoka Lake near Lubbock. Among the discoveries are the remains of a *pastores* sheep corral. (Courtesy of Lubbock Lake Landmark; Jim Gannaway, photographer)

a rock corral in Yellowhouse Canyon at Buffalo Springs. The site can no longer be found, and Hicks and Johnson speculate that it may have been engulfed when Buffalo Springs Lake, a man-made lake east of Lubbock, opened in 1960, or that the corral was dismantled after 1932.[13]

Other *pastores* sites are located in Blanco Canyon, southeast of Floydada, and just south of the mouth of Yellowhouse Canyon.[14] Yellowhouse Canyon begins in Lubbock and stretches about thirty miles until it ends downstream from Southland, east of Lubbock. The canyon starts where Yellowhouse Draw and Blackwater Draw meet in Mackenzie Park to form the North Fork of the Double Mountain Fork of the Brazos River.

While it was unusual for *pastores* to have come as far south along the eastern edge of the Llano Estacado, one *pastor*, Jesús Perea, moved his sheep down to Tahoka Lake, south of Lubbock. Whether Perea used the other sites is speculative. Large cattle ranchers also found the Panhandle and South Plains grasslands favorable, and they moved their herds into the region. By 1881, titles to vast acres were granted to the cattlemen, and the Texas legislature enacted a barbed-wire fencing law. In 1884, another law required that state lands must be leased prior to any use as pastures. Faced with a combination of these laws and several harsh winters in a row, the *pastores* were forced out of the region and back to the Territory of New Mexico by 1887.[15] The *pastores* left behind infrastructure that the early ranchers later used.

Mrs. May is proud that the Quanah Parker Trail organization recognizes her land as part of the Quanah Parker Trail. Parker, who took his mother's Anglo name, was the last chief of the Kwahadi Comanche. His grandson, Baldwin, visited Mrs. May several years ago to confirm that the site should receive one of the dozens of 22-foot-tall steel arrows that mark areas where the Comanche and Quanah Parker lived, traded, and fought. The son of Comanche Chief Peta Nocona and Cynthia Ann Parker, a white woman captured by the Comanches, Quanah resisted pressure to move his band to reservations in Oklahoma, mounting an unsuccessful war against Anglo expansion in Northwest Texas.[16]

A historical marker located near Snyder, Texas, notes that the Comanche named Quanah Parker chief on the reservation. It states that he was an able leader, negotiating grazing rights with Texas ranchers, investing in businesses, and lobbying the US Congress on behalf of the Comanche Nation. Parker, who died in 1911, was often a hunting partner of Theodore Roosevelt.[17]

While Mrs. May is happy to tell visitors about the lake's archaeological and anthropological significance, she is perhaps most excited by the wildlife and beautiful wildflowers around the lake, especially the Tahoka daisy. The flower was first noticed by Jack Alley's wife, blooming around the lake, in about 1898. It is a low-growing plant with fernlike leaves and bright purple flowers radiating from a yellow center.

Mrs. May's most fervent wish is that the land is safeguarded for future generations. "I just hope that the people who come after will preserve it and have it for the same reasons I do, for education, ecotourism, and historical interest."

CHAPTER 7

FIRE AND FLOWERS

Ask people unfamiliar with West Texas to define the Llano Estacado and you will most likely get words like hot, dry, and shades of brown. While those terms are accurate, the region is not a desert but has been a grassland for millions of years.

Although humans first appeared on the Llano Estacado at least 13,000 years ago, the first written description did not occur until 1541, when Spanish explorer Francisco Vázquez de Coronado crossed the region. In a letter that year to the King of Spain, Coronado wrote:

> I reached some plains so vast, that I did not find their limit anywhere I went, although I traveled over them for more than three hundred leagues [nine hundred miles] . . . with no more landmarks than if we had been swallowed up by the sea. . . . [T]here was not a stone, nor bit of rising ground, nor a tree, nor a shrub, nor anything to go by.[1]

Coronado also wrote that while he found excellent pastureland, he was disturbed by the endless prairie, with no markers and limited water.

Ironically, Coronado's sea of grass with no markers to tell where he was is now known as the Staked Plains. From Spanish to English, Llano translates as plain or prairie; Estacado translates as "to tie to a post." Llano Estacado references the stakes or posts driven into the plains by early settlers to mark trails and water resources. Others, such as Art Leatherwood, who writes in the *Handbook of Texas*, say that palisaded plains might be more accurate. A palisade indicates a fence of stakes, usually for protection or as a line of bold cliffs. In his diary of an 1841 expedition from Texas to Santa Fe, explorer Thomas Falconer described the steep escarpments as "elevated or palisaded much as palisaded sides of a fort."[2]

EARLY SETTLERS

Other early explorers found the Great Plains to be as inhospitable as had Coronado. Edwin James, a member of Major Stephen H. Long's 1820 expedition to explore the Rocky Mountains, wrote, "Although tracts of fertile land considerably extensive are occasionally to be met with, yet the scarcity of wood and water, almost uniformly prevalent, will prove an insurmountable obstacle in settling the country."[3] Long, an army engineer, produced an 1823 report that featured the first map of the Great Plains to label the area the Great Desert. The Llano Estacado lies toward the southern end of the Great Plains, which ends at the Edwards Plateau around Central Texas.

Others would have strikingly different views. Dr. J. W. Hunt and his family were among the early settlers on the Llano Estacado, arriving in June 1881. The Hunts were part of a group of Quakers who arrived at a Quaker colony founded by Paris Cox at Estacado, near what would become Lubbock. Hunt would write, "As we swung around the final obstruction of the rock cliff and mounted the final ascent, there suddenly opened before our wondering eyes the mighty stretch of the Great Plains of Texas."[4] He wondered at the lush grasslands and the brilliantly colored wildflowers. He closed his description with, "The landscape dotted with little lakes gleaming in the morning sun seemed a vast emerald shield embossed with gems of silver."[5]

While the early notations of fertile land and lush pastureland were prophetic, the thought that the land would be difficult, if not impossible, to settle, and Long's desert observation proved wrong. Hunt reported that the first crops ever planted on the Llano Estacado in 1880 included corn, oats, sorghum, and potatoes. He noted that all did well. Hearing of the successful crops, other Quaker families began moving to West Texas. The third family to arrive at Cox's colony was George Singer's.[6]

Crops continued to thrive on the South Plains, and today the area is one of the world's largest cotton producers. Along with other crops such as wine grapes, sorghum, and wheat, cotton forms the foundation of the region's agricultural economy.

The *pastores* agreed with Hunt's assessment of a vast grassland with plentiful water. However, Johnson points out that the *pastores* had a sentiment about the land that was a bit different than that of early settlers looking for a new place to live. Johnson notes in her chapter in *Archaeological Landscapes on the High Plains* that the *pastores* called the western edge of the Llano Estacado *la ceja de Dios*, the eyebrow of God.[7]

DEVELOPMENT OF THE LLANO ESTACADO

Settlers began developing the Llano Estacado in 1870. The US government had forced Native Americans onto reservations around 1851, and the massive bison herds were all but eradicated, leaving the Llano Estacado accessible to settlers, particularly ranchers who took advantage of the vast open grassland. The *Handbook of Texas* notes that the Llano Estacado became the fastest-growing part of Texas, "progressing from an economy based on unfenced public grazing land to a modern industrial economy within half a century."[8] Around 1920, the changing economy, rising costs, an increased demand for land by new settlers, and the growth of farming led many large ranchers to sell off portions of their land. Even with limited water sources, farmers flourished by developing new pumping methods to draw water from the Ogallala Aquifer for crop irrigation. While farming and crop-watering techniques have changed over the decades to lessen the drain on the aquifer, irrigation and city growth have seriously depleted the Ogallala.[9]

About ninety-five percent of the state's almost 172 million acres are privately owned today. Most, about 57 percent, are in the hands of ranchers and farmers.[10] The massive King Ranch in South Texas encompasses approximately 825,000 acres, making it larger than Rhode Island. The remaining 38 percent of privately owned acreage is forestland controlled by families, industry, and state and local governments.[11]

PRAIRIE RESTORATION

With public access to the prairie vanishing, Eileen Johnson has fought to restore the natural habitat at the Landmark. The 336 acres include land not under excavation but restored as close as possible to its natural state of the late 1800s. Johnson faced the challenge of achieving this goal while not affecting the archaeological deposits that she is sure exist on the unexcavated land.

While preserving the natural history of the Landmark is the overriding objective, Johnson has a personal appreciation for the area. Even though she grew up in California, with its mountains on one side and the ocean on the other, the prairie is where she feels at home. "Coming to the Great Plains felt like where I was supposed to be," she says. "Even doing research in Argentina, I got off the plane and thought, wait, did I leave Texas? This looks just like Texas. The pampas of South America look just like the prairie of Texas and the rest of the Great Plains."

Figure 34. When Eileen Johnson took over at the Landmark, the draw was overrun with mesquite and other brush. Beginning in 2000, Johnson set in motion a plan to restore the prairie to its natural state. (Courtesy of Lubbock Lake Landmark; Kippra D. Hopper, photographer)

By 2000, Johnson had a plan to reclaim the Landmark's prairie. "The natural history out here is unparalleled," she says. "We have native grasses, small native animals, and a good bird population. I knew it would take a lot of effort. I needed someone who understood the importance of both the Landmark and the prairie and what it would take to bring the prairie back so the land could heal."

Johnson found the perfect partner in Sterling Scott Trevey. With a degree in agronomy from Lubbock Christian University, Trevey brought years of experience to the job. He grew up and worked for several years on a ranch, gaining experience in various conservation techniques. "Any time ranching was involved, it just lit me up," he smiles. "I love the livestock; I love the country. Growing up, I wanted to work with the land and livestock. I am not involved with animals except for the wildlife, but I love the land here."

Determining what the prairie looked like took a bit of detective work. Trevey and others at the Landmark pored over historic photographs to try to see the land as it had been. Trevey was particularly drawn to the work of Erwin Smith, who documented ranch life between 1905 and 1920. The Amon Carter Museum in Fort Worth holds an extensive collection of his photographs. "By blowing up the images, we could get a rough idea of what the land looked like through those photos," Trevey explains. "We could see what type of brush was on the land and how much there was. We could see if there were sunflowers present. Though Smith's images are a bit later than we were shooting for, it is similar."

Trevey's job description is broad: care for the prairie, keep the equipment and vehicles running, and fix and/or build things when needed. He knew how to do some of those chores coming in; some he had to learn as he went. His most significant challenges were clearing out the mesquite that had taken over the area and repairing the damage to the land from various misuses, especially from off-road vehicles over the years, while keeping the archaeology intact.

It was a daunting task for even the best equipped of places. Trevey, however, quickly realized that all he had to work with was a power drill and a sparsely equipped toolbox. Johnson laughs when the topic comes up, saying he had more than that: There was also a hoe and an old lawn mower from her garage. "I remember the first things I bought were some shovels, loppers, and a double-headed ax," Trevey says, shaking his head. Texas Parks and Wildlife Department took its maintenance equipment to its new regional headquarters a few blocks away when the Landmark and TPWD dissolved their partnership in 2000.

First on a very long list of things to do was to get the brush that had taken over the land under control. To do that, Trevey had to find a suitable herbicide that would kill the plants but not affect wildlife or, most importantly, organic materials such as charcoal that might be sent for radiocarbon dating in the future. Dr. Johnson and Trevey drew upon an ongoing relationship with the University of Arizona to help them choose the correct herbicides. The University of Arizona was already performing radiocarbon dating for the Landmark. "The folks in Arizona told me if it were water soluble, we'd be ok; if it was slightly soluble, then it could affect the process; if it were insoluble, we'd need to avoid that," Trevey says.

For the mesquite, Trevey settled on a low-volume liquid herbicide. It is applied with a spot injector gun, similar to a syringe, that measures in increments of milliliters (i.e., cubic centimeters). Trevey lops off the mesquite near ground level, then injects the herbicide into the ground; rainwater carries it to the plant's roots. This hand-application method is targeted and labor-intensive. However, Johnson notes, "It was necessary not to disturb the ground or kill off non-targeted plants. While rooting/chaining the mesquite trees using plows to pull up the trees and then using chains to smooth the ground is a standard method, it greatly disturbs the ground and potential archaeological deposits." Other plants required different herbicides that could be sprayed on the leaves.

Native to and ubiquitous on the semi-arid South Plains, mesquite loves the hot, dry conditions of the region. The oldest record of the tree at the Landmark/Yellowhouse Draw system dates back eight thousand years. The plant does have benefits. Its dry wood burns hot, producing an aroma that makes it popular for smoking meats. Various Native American tribes on the Llano Estacado favored mesquite wood for use in their hearths. Johnson says they have recovered mesquite charcoal in hearths ranging from San Jon, New Mexico, to Stanton, in the Permian Basin, from Post to Tule Canyon, and west to Saddleback Mesa around Tascosa, near Amarillo.

Not only is it a favored cooking method, but over the centuries, mesquite wood has been used to build houses, fences, and tools;

its sap is used in medicinal salves; and its beans are a food source for humans and animals. Its downside is its ability to overtake the land where fire has been suppressed. The trees can survive droughts, and their pods produce a vast number of seeds that can remain viable for a decade.[12]

The mesquite and grasslands consume about the same amount of water during a rainy season. However, the grasses go dormant during a drought while the trees continue to suck up water. The roots run so deep that they can access water the grasses cannot, reducing the amount of water available to meet human and animal needs.

By 2002, his second year on the job, Trevey and his crew had hauled away about seventy tons of brush. By the end of April 2005, approximately 53.6 acres of mesquite and five acres of Siberian elm trees had been controlled. As part of a beautification project, the fire department had planted Siberian elm, a non-native species, around Lubbock Lake in 1938. These trees are far more invasive and detrimental than mesquite and can cause erosion due to their deep root systems; also, much like mesquite, they will force out more desirable plants.[13] Trevey and his team used a large woodchipper to turn the trees into several 20-foot-high piles of mulch. They've used the mulch over the years for projects around the Landmark.

Over time, most of the mesquite has been removed, leaving only a few trees. "There were a few mesquite trees that were large enough and old enough, maybe around one hundred years old, that we left them," Trevey explains. "Mesquite would have been present on the prairie in the late 1800s." Mesquite trees have grown in the draws and canyons for thousands of years. Johnson emphasizes that the prairie restoration was never meant to eliminate the native mesquite but rather to bring it under control.

Trevey has seen the re-emergence of some native trees since the mesquite and Siberian elm were contained. He has planted some native trees, such as the cottonwood and Texas walnut; others have sprung up independently. Johnson explains that at the Landmark, "native" means trees and wildflowers native to the Llano Estacado and present in the Landmark's fossil record or seed bank. Some trees, grasses, and wildflowers, such as bluebells, may be native to Texas, but not necessarily to the Llano Estacado. Trevey notes that the netleaf hackberry and western soapberry are among the trees that have flourished. While he would like to remove more Siberian elms, he wants the mature trees to remain as habitat for Mississippi kites, golden eagles, and other birds visiting the Landmark.

Not all the damage to the grassland is due to invasive plants; it also results from human use. Off-road vehicles marred several acres of land, potentially damaging artifacts on or near the surface and chewing up the soil, creating deep ruts. That torn-up land eventually became a thick stand of tumbleweeds, another non-native invasive plant that can injure the ecosystem.

Native to Eurasia, the tumbleweed, or Russian thistle, has become common in North America. This is an annual plant that breaks off at its base when it dies. When the wind blows, the brown, thorny plants tumble across the ground and spread their seeds as they roll. The large, ball-like plants can roll across highways, causing traffic hazards. They also tend to pile up against houses and other structures, creating a sizeable thorny mass; because they are dry, they pose a fire hazard.

Despite the strong foothold the tumbleweeds had at the Landmark, Trevey found they were easier to control than the mesquite. He learned that mowing the weeds as they produced blooms killed them, allowing native grasses to flourish. Contrary to some theories, tumbleweed did not come to North America from cattle and other shipments from Mexico. It first appeared in South Dakota in the 1870s, having hitched rides in sacks of grain from Russia.[14]

The Landmark's prairie is mostly restored today; however, Trevey knows the job will never be finished. Johnson tells the story of Trevey storming into her office one day, upset over his latest

Figure 35. Tumbleweeds have been mostly controlled at the Landmark, but West Texas winds can blow the thorny plants across the prairie to catch on fences, just as they do in city neighborhoods. (Courtesy of Lubbock Lake Landmark; Kippra D. Hopper, photographer)

challenge. "'We have nut grass! Nut grass!' He was really upset. He takes these things personally," she laughs. His latest nemesis is yucca. The yucca is not a cactus but is a succulent that is part of the agave family. Trevey is using an herbicide to control the plants, not eliminate them. Yucca thrived on the Llano Estacado grasslands as far back as 26,000 years ago.

Broomweed is another continuing problem that Trevey keeps a close eye on. Broomweed belongs to the sunflower family and is a warm-season native plant. Trevey says that before he arrived at the Landmark, the voracious occupants of a large prairie dog town in the northeast section of the site had eaten most of the grass, allowing broomweed to take over. Texas Tech range experts had hypothesized that the broomweed might be toxic to the grass and other plants at the time.

Trevey notes that newer research has shown that's not the case. However, broomweed is still damaging because it sucks the ground moisture away from the grass. Once the broomweed was removed, the grass and other plants thrived. Trevey, for whom statistics are essential markers of the prairie's success, observes that the amount of grass went from fifty pounds per acre in 2005 to about 1,500 pounds today.

A 2017–2018 exhibition at the Landmark's Nash Interpretive Center explained the underlying philosophy of the Landmark's restoration efforts: Use nature's methods when possible, imitate them as needed, and recreate conditions that would allow the grassland to heal itself and flourish.[15]

Following that philosophy, Trevey has found a way to mimic grazing, an integral part of land management best practices, if controlled. Trevey notes that overgrazing rangeland is detrimental to the land and leads to reduced ground cover, which in turn allows more erosion, reduces the soil nutrients that plants need to grow, and makes it harder for water to penetrate the ground. While no animals are grazing on the Landmark's prairie, to mimic good grazing practices the Landmark team will shred or mow the grassland, removing the top layer of grass, which generates new growth.

Trevey's alternative to grazing? Fire. He explains that fire is nature's original method of restoring the land. For thousands of years, Native Americans used fire to revitalize their land, improve habitat for wildlife, and enhance grazing for bison. While researching the history of fire use, Trevey came across an extensive website developed by the Confederated Salish and Kootenai Tribes of the Flathead Reservation. Headquartered in Pablo, Montana, the confederation consists of three tribes: the Salish, the Pend d'Oreille, and the Ksanka Kootenai. They created the Fire on the Land website to educate the general public about the tribes' historical and current use of fire. The site features historical photos and interviews with tribal elders detailing how fire shaped the land, how fire suppression altered the land, and how the reintroduction of fire is restoring the prairie.[16]

Although the Landmark had been conducting controlled burns for several years, Trevey says he is always interested in furthering his knowledge about fire. What drew his interest was the tribes' ability to trace the effects of fire on their land back to the 1600s. In 2015, Trevey traveled to Montana to visit the tribes' fire experts. "I was able to spend some time with those folks," he says. "They took me through their land and showed me cross-sections of thousand-year-old ponderosa pines that show signs of fire dating back to 1650. They told me that when their people got ready to leave an area, they would set it on fire, knowing that when they returned in a few years, the native grassland would be new and fresh."

For more than a century, policies aimed at suppressing wildfires have had the unintended consequence of allowing more fuel to accumulate, making wildfires harder to control. When used correctly, fire is a natural part of the ecosystem and is highly beneficial to the land.[17]

Smokey Bear remains one of the most successful advertising campaigns ever designed. Launched in 1944, this USDA Forest Service initiative is credited with raising awareness about forest fires. However, some fire and forestry experts say it might have been too successful. In a 2007 article published in the journal *Frontiers in Ecology and the Environment*, Geoffrey H. Donovan and Thomas C. Brown, USDA Forest Service staff members, argue that Smokey's message contributed to decades of aggressive fire suppression. They conclude that a century of wildfire exclusion "has substantially altered many of the nation's forests and resulted in wildfires that are more expensive and more difficult to control." The authors called for a fundamental shift in federal fire policy, one that would require the public to change its expectations of wildfire suppression. The authors write, in part, "There must be recognition that complete wildfire exclusion is neither desirable nor possible and that maintaining forest health and controlling suppression expenditures necessitates burning large areas of forested land annually."[18]

Figure 36. An important part of managing the Landmark's prairie land is prescribed burning. Fire can revitalize the land and improve habitat for wildlife and plants. (Courtesy of Lubbock Lake Landmark; Stance Hurst, photographer)

Today, the Smokey Bear website continues to promote wildfire education and fire prevention techniques, offering educational materials for kids and educators. The campaign emphasizes the prevention of unwanted and unplanned fires while supporting prescribed burning.[19]

Keith Blair, who owns Red Buffalo, LLC, a prescribed burn company, and who works with the Lubbock Lake Landmark to manage its prairie, agrees that the Smokey campaign unintentionally went too far. At the same time, he applauds its fire education efforts and would like to take them further. He has pushed for an elementary school fire safety and prevention curriculum.

A key element in controlling wildfires is for landowners to remove brush and mesquite trees that can fuel an unwanted or uncontrolled burn. Blair tells the story of a wildfire near Aspermont, Texas, from a few years ago. "Part of that fire ran into an area we had previously burned," he explains. "There was no fuel, and the fire was out. That allowed firefighters to move on to other areas."

The art and science of prescribed burning is partly derived from Texas Tech University Professor Henry Wright's research. Wright is considered one of the most influential people in the prescribed burning world. His research, classroom education, and workshops helped establish fire as an acceptable ecological tool to control brush and weeds on rangeland and to restore grasslands.[20]

A certified burn manager, Blair earned his bachelor's and master's degrees at Texas Tech and studied under Dr. Wright and Professor Carlton Britton, who followed in Wright's footsteps as a leader in the science of prescribed burning. Blair and Trevey have formed a partnership of sorts over the years. The Landmark offers Blair a grassland to train others in burn management, which spares the Landmark the expense of maintaining its burn program. Trevey has completed all the requirements to become a burn manager and helps Blair with burns at the Landmark and other areas around the region.

Burning is expensive, running around $3,000 per day or more. That cost is in addition to the $2 million insurance policy the burn manager or landowner must carry. Blair explains that the expense includes transportation, paying his crew, and equipment, such as fire trucks and all-terrain vehicles used to monitor the fire. His crew may travel four to five hours to get to the burn location. "Our trucks get only about eight miles to the gallon," he says. Blair's company does ten to twenty burns annually, depending on demand and weather conditions.

Blair was a leader in establishing the Texas Prescribed Burning Board in 1998. Today, under the auspices of the Texas Department of Agriculture, the board sets standards for prescribed burning and develops training for certified burn managers. It comprises state agencies, higher education institutions (including Texas Tech), and private landowners.[21]

Burn managers follow a precise set of criteria. Blair credits Wright with developing the standards that are still used today. Blair used what he learned from Wright and Britton to create the 60:40 rule: The air temperature must be below 60 degrees Fahrenheit but not less than 40 degrees Fahrenheit; the relative humidity must be greater than 40 percent but less than 60 percent; and wind speeds cannot exceed eight miles an hour at eye level. Trevey emphasizes that a burn should only occur if these three points are met.

The 60:40 rule is used when creating a black line, the first step in any burn. The black line is designed to help safely manage the primary head fire by stopping it. Blair explains that a head fire is used to achieve a range of objectives such as removing heavy brush.

Given the small size of the Landmark, at just more than 336 acres, only about eighty acres of the prairie are burned at any one time. Trevey notes that burning at the right time of year helps kill invasive species and can ultimately lead to less use of herbicides, which is good for the ecosystem. Trevey began using prescribed burning at the Landmark in 2007 and marvels at how resilient the land is to fire. A rainy spring helps the grass and wildflowers return quickly, providing improved habitat and food sources for the wildlife. He remembers an eighty-acre burn in December 2010.

"We had had above average rainfall that year, and it was just lush here, so there was lots of fuel," he says.

As happens on the South Plains, a rainy year is often followed by a dry year. "About two weeks after that 2010 burn, we had a massive haboob that just swept the ash from the land," Trevey remembers. A haboob is an intense and enormous dust storm that can produce a massive wall of dirt as it moves across the land. This one was powered by winds of almost seventy miles per hour. The wind and dirt scrubbed the land clean, leaving only grass stubble no more than three centimeters tall.

Exploring the area in August or September of 2011, Trevey found almost nothing growing. He says mesquite, southwestern prickly poppy, and buffalo gourds were the only green plants. "It looked like the surface of the moon," he says, shaking his head. "I was afraid we had killed it; there was nothing green. I thought we would have to reseed the area." Rather than reseed right away, Trevey decided to wait until the following year, which turned out to be a wise decision.

After about eight inches of rainfall in the spring of 2012, Trevey began to see signs of blue grama grass, the dominant grass along the upper edge of the bottom of the draw. "The seeds were in the soil and were just waiting for rain," he says. "By 2012, the blue grama was back, and a couple of years later, we were back to the full diversity of grasses and flowers. It was just amazing."

Trevey and his crew have used limited native grass and wildflower seeding across parts of the prairie to heal past damage, preferring to let the land grow naturally. After twenty years of closely studying the grassland, Trevey believes that the topography of Yellowhouse Draw influences what and where things grow. The draw consists of four levels: the axis, the lowest point in the draw; the margin, the slope between the axis and the beginning of the wall; the wall, the rock edge just below the Caprock; and the upland or rim, which is the highest point of the draw. Trevey

Figure 37. Because of careful land management practices, the Texas horned lizard—once plentiful in the arid and semiarid parts of the US but now listed as a threatened species in Texas—flourishes at the Landmark. (Courtesy of Lubbock Lake Landmark; Kippra D. Hopper, photographer)

explains that different plants grow in specific areas based on the soil mixture. "The plant diversity we get is pretty unique, I think," he says. "The one thing I believe is if we're taking care of the land, everything else, the plants and wildlife, will take care of itself."

Trevey points to the Texas horned lizard, or the horny toad as it is popularly known, as a beneficiary example of conservation efforts. The lizard is a scary-looking little critter that is recognizable for the multiple horns on its head, including two central longer horns. The small reptile was once plentiful in the arid and semi-arid parts of the country but is now listed as a threatened species in Texas. The primary reasons for their decline are the destruction of their habitat and reduced available food. However, visitors walking around the Landmark are apt to see the little creatures regularly. The reason, according to Trevey, is good land management practices. "A lot of the places that are not seeing the horned lizard are doing very little, if any, land management," he emphasizes. "Proper land management results in more cover and food sources that increase horned lizard populations."

The Landmark has been home to mammals for millions of years. The mammals' remains are essential to the archaeological record as they provide environmental clues. For example, bison have been Lubbock Lake's primary game animal for humans for 13,000 years. Their presence indicates that the area was an extensive grassland.

Restoring the prairie has led to an increase in wildlife at the Landmark. Trevey notes that the only mammals he saw when he first started his work there were jackrabbits and cottontails. The prairie was so choked with mesquite and other brush that many animals could not make good homes. Now, visitors encounter a wide array of animals. The jackrabbits and cottontails still call the Landmark home, as do dozens of other animals, such as deer, squirrels, mice, gray and red foxes, and coyotes.

Prairie dogs, which have inhabited the western part of the state for well over one million years, are partial to short grass prairies. They are social animals and live in what can grow into large colonies of burrows connected by underground tunnels. While many think the little animals are cute and fun to watch, others, including city developers, consider them a nuisance and a danger to livestock. Prairie dogs have faced widespread extermination efforts over the decades. According to the World Wildlife Fund (WWF), "Today, prairie dogs occupy only about 2 percent of their historic habitat."[22]

Prairie dogs are a significant part of the grassland ecosystem. They are considered a keystone species, meaning other animals depend upon them for food or shelter. The WWF notes that the animals support about 136 other species. They are food for several different animals and birds. Abandoned burrows are used by the burrowing owl, the horned lizard, and rabbits, among others. Removing a keystone species can damage the ecosystem.[23]

Ranchers have eliminated prairie dogs because they thought the creatures competed with cattle for grass. There is also a common

Figure 38. Prairie dogs are an essential part of the grassland ecosystem. They are humanely managed at the Landmark to protect buried bones and artifacts from their tunneling. (Courtesy of Lubbock Lake Landmark; Kippra D. Hopper, photographer)

misconception that livestock can step into a prairie dog's burrow and break a leg. Johnson defends the animals, explaining that the creatures have coexisted for thousands of years with bison. "Bison are a herd herbivore related to cattle," she says. "I read a scholarly publication recently that said cattle stepping in prairie dog burrows and breaking a leg is folklore. Much like mesquite, the prairie dogs are part of the natural environment and ecosystem."

Studies in recent years have indicated that cattle prefer to graze among prairie dog colonies because of the superior grass quality. One study by the Borderlands Research Institute for Natural Resource Management notes that the grass amounts were similar on and off prairie dog colonies and that the protein levels of the grass growing on prairie dog colonies are 10 percent higher.[24] The little mammals flourish at the Landmark, but because their burrows can damage the archaeological deposits, experts humanely relocate them to places such as Caprock Canyons, which, Johnson notes, welcomes prairie dogs.

While the wildlife has flourished, so too has the flora.

Trevey regrets not doing a more thorough survey of the grasses when he first started, so he could compare them to what exists today. "We had a few types when I started in 2000, but the mesquite, tumbleweeds, and other brush choked out much of the grass," he says. "Today, we have tons of different types of grass and wildflowers."

Thanks to Daria McKelvey, who studied botany as an undergrad at the University of Texas, Trevey knows there are around two hundred species of native wildflowers, cacti, grasses, and shrubs. McKelvey wanted to learn about the native plants of her home in Lubbock. She noted that there was no class in plant identification at UT, so she began to walk the Landmark's trails when she was home in Lubbock and learned to identify the different plants.

Because many wildflowers are native to the Llano Estacado, Trevey wonders why they were not mentioned in the early writings of Coronado and other explorers. "Maybe the flowers weren't in bloom, or it was a year when the moisture wasn't conducive to many flowers," he theorizes. "Or maybe they could not be seen for the 'sea of grass.'" Relevantly, in a summer 2020 edition of the Landmark's newsletter, McKelvey wrote, "The typical stereotype is one of a flat, monotonous, and uninteresting landscape, when in fact, the region showcases a magnificently dynamic terrain with striking floral displays unique to the area."[25]

McKelvey, who earned her master's degree in horticulture at Texas Tech and now supervises the Kemper Center for Home Gardening at the Missouri Botanical Garden, echoes Trevey's admiration for the plants' survival instincts as they manage to thrive in the hot, dry conditions of the Llano Estacado.

McKelvey believes there should be more efforts to educate people about the beauty of the Llano Estacado and get them to recognize its diversity of plant life. To do that, McKelvey says those promoting the region must change the narrative to focus on the variety of plants, their resiliency, and the beauty of the Llano Estacado. "I do not believe that an area has to be green to be beautiful," she insists.

When there is sufficient moisture in the fall and winter, the Landmark comes alive in spring and summer with brilliant wildflower colors, ranging from yellow to red and orange, pink to blue, and purple to white. The American basket flowers (*Centaurea americana*) are among the showier of the Landmark's wildflowers, and during a good year, they blanket the draw in purple and white. The flower is native to the American Southwest and has no other human use except as decorative. Its seeds, however, are a winter food source for dove and quail, both frequent visitors to the Landmark.

Native Americans used many wildflowers. The Indian blanket (*Gaillardia pulchella*) is a brightly colored wildflower with a red center and yellow edge. It thrives in poor soils, and Native Americans used parts of the plant to treat headaches. The native copper mallow (*Sphaeralcea angustifolia*) is one of about one thousand species of the mallow family, including hollyhocks, cotton, and okra. The European marsh mallow species is today's marshmallows' original ingredient. Native Americans made the plant into poultices to relieve pain. The buffalo bur (*Solanum rostratum*) is among the oldest plants at the Landmark, dating back about 10,000 years. The name comes from a belief that the thorny seeds spread by attaching to bison coats. While it is poisonous to livestock, some wildlife and birds feed on its seeds, although its thorns mean it is not a favorite food source. It is safe to grow at the Landmark because there are no livestock.

Among the dozens of flower varieties are sunflowers and several types of daisies, including the Tahoka daisy. There are chocolate daisies, which really do smell like chocolate, and the hog potato flower smells like Fruit Loops. "That's not something you can get from a textbook," McKelvey laughs.

McKelvey and Kippra D. Hopper, a published photographer and part-time employee of the Landmark, agree that the ability to smell the flowers and see their different textures and colors makes the Landmark special. Hopper is particularly drawn to the wildflowers and has learned things she never expected. "I've discovered that the wildflowers don't like rich soil," she says. "They like hardscrabble soil. I know that is counterintuitive, but they like growing where the caliche is thick. You will see them seemingly hanging off the sides of outcroppings." Hopper also notes that many native plants are critical to bees and other pollinators, with some plants drawing specific pollinators to the Landmark.

Hopper and McKelvey are certified as Texas Master Naturalists, a volunteer group that has become important to the Landmark. Hopper, a past president of the South Plains Chapter of the Texas Master Naturalists, explains that to become a member, one must complete forty hours of training, forty hours of volunteer work, and eight hours of advanced training within twelve months to become certified. To continue their certification, members must undergo eight hours of advanced training and do forty volunteer hours yearly. The group does everything from leading tours to picking up trash at the Landmark.

For Hopper, the Landmark defines the phrase "sense of place." Hopper explains that, for her, a sense of place means writing about your home and what is around you, opening your eyes, and seeing. "You see a sense of place used in literature," she says. "I wish people would realize just how beautiful it is here. Getting out into nature gives you a strong connection to your ecosystem."

The prairie gives the Landmark another way to educate people and allow them to experience nature without leaving the city. The Landmark crews have built four-and-a-half miles of trails to allow the public access to as much of the area as possible.

The longest trail, the Nature Trail, covers three-and-a-half miles of rolling terrain, allowing visitors to explore native vegetation and observe birds and small animals in their native habitat. Its natural slopes challenge walkers and runners, with several high school and college track teams training on the trail. It is also popular with bicyclists. There is also a two-and-a-half-mile alternative of this path. The routes have interpretive panels to help visitors learn about the landscape. The trails are kept graded, but their surface is dirt and caliche.

Both versions of the trail offer a view of an ancient playa lake that sits inside the loop. Playas, the Spanish word for beach, go through wet-dry cycles today; however, during the Pleistocene and early Holocene, they held water year-round. The wet/dry cycle most likely occurred during the middle Holocene drought, which lasted at least two thousand years.[26] Rain fills the somewhat circular wetlands. The playas provide water sources for animals during wet periods. However, the Landmark's playa has so much silt blown in from the prevailing southwesterly winds that it rarely contains water; the rain drains through the silt. On the outside of the trail is a large dune on its northeast side, formed by the wind blowing sediment from the dry playa. It is estimated that the dune began forming about 30,000 years ago.[27]

The half-mile Archaeological Trail takes visitors back 13,000 years as they stroll around the Landmark's research compound. The trail features interpretive panels to help visitors understand the excavations. During the July–August dig season, visitors can stop and watch an active excavation. Groups can arrange guided tours through the Landmark offices.

The Llano Estacado Wildflower Trail is ADA-accessible, while the others are not. The trail features a boardwalk to help with mobility issues and frequent shade structures and benches. Signs along the path identify many plants and trees and recount some of the Landmark's nineteenth- and twentieth-century history. The Wildflower Trail allows everyone to get out into nature, which Deborah Bigness, the Landmark's former operations manager, says makes the trail different from most. "When you go to a state or national park, or even a city park, they mostly get you to the edge of nature," she explains. "Our trail aims to get everyone into nature to see that flower up close or that little animal, assuming it will hold still."

The Wildflower Trail is a feat of creative engineering. Trevey's team built the trail, shade structures, and benches without digging one hole so that the ground and any potential artifacts lying underneath were not disturbed. After puzzling over how to accomplish that task, the Landmark crew finally decided to construct the trail as if it were portable. The solution: laying concrete blocks and covering them with what would become the frame for the trail. Four-inch by four-inch by eight-foot beams were anchored to the concrete beams, and then Trex decking was screwed into the beams. The shade structures, called sails because of their triangular shape, are anchored in concrete buckets. The interpretive signs explaining the natural history are also movable, allowing staff to change the explanations if a different flower or grass appears.

The project took eight years to complete, opening to the public in 2009. Trevey notes that he and his assistants, when he had help, drilled 270 three-inch screws into the boardwalk beams every eight feet, burning up one drill. The sails and track have survived years of strong West Texas winds, thunderstorms, and occasional ice and snow with little to no damage or having to be moved for any reason, until vandals destroyed part of the trail in 2025. As of 2026, efforts are underway to raise the funds needed to rebuild the trail. "This is an important project in that it helps the public understand the significance of this place," says Bigness. "The Landmark takes imagination on the part of the visitor. We are asking people to ignore the telephone poles, the fences and buildings, and the sound of trains going by and imagine what this land was like before humans arrived."

All trails are accessible only during the Landmark's regular operating hours. During the spring and summer, however, the Night

Hike program allows walkers to experience the Landmark after dark. Night-blooming wildflowers and more nocturnal wildlife are visible. Periodic Sunrise Saturday Guided Hikes are also offered.

The prairie restoration honors both the more modern prairie and wildlife and the ancient animals that roamed the draw millions of years ago. During the Landmark–Texas Parks and Wildlife Department partnership, the Lubbock Lake Site Foundation commissioned four sculptures representing Ice Age animals. The Foundation funded the first three, and TPWD covered the costs of the final sculpture, that of the Columbian mammoth.

The committee overseeing the sculpture project comprised Johnson, former Museum of Texas Tech director Gary Edson, and civic leaders. They were charged with selecting which animals would be depicted in the bronzes; creating specifications for each animal based on bones discovered at the Landmark; raising the funding; and requesting applications from artists nationwide to execute the bronzes. Each artist submitted a model and their credentials to the committee.[28]

The first bronze was the giant short-faced bear created by artist Lisa Perry. The living animal stood eleven to twelve feet tall on its hind legs. Weighing in at as much as 1,500 pounds, the bear was twice the size of the modern grizzly. Scientists estimate that despite its immense size, the bear ran thirty to forty-five miles per hour. Evidence also suggests it was a scavenger rather than a hunter. Organizers dedicated Perry's bronze in 1993.

The following year, the ancient bison sculpture, also produced by Lisa Perry, was dedicated. *Bison antiquus* was a massive animal, about a third bigger than the modern bison. It was the most common large plant-eating animal in North America. The immense grassland of the Great Plains was perfect for the bison. It is one of the few large animals of the Ice Age that did not go extinct; instead, it adapted and became smaller.

The giant pampathere sculpture by David Thomas was dedicated in 1995. Commonly called a giant armadillo, it looks much like its modern cousin, the nine-banded armadillo, although much larger. Local lore maintains that the creature's statue, which reflects the dimensions of the living animal, is the size of an old Volkswagen Beetle. In truth, the giant pampathere stood only about three feet tall, was six feet long, and weighed more than five hundred pounds. While that is big, it is not Volkswagen big. The pampathere is Johnson's favorite sculpture because it is kid sized. Children are encouraged to sit on the bronze; its more accessible height makes it an excellent background for photographs.

The Landmark and TPWD dedicated the final bronze in 1999. TPWD Director Andrew Sansom says the Columbian mammoth with her calf sculpture was funded by TPWD and created by TPWD employee Michael O'Brien. Standing fourteen feet at the shoulder, the Columbian mammoth weighed between eight and ten tons. The Columbian mammoth was sleeker and larger than its more familiar cousin, the woolly mammoth, which was primarily found in colder northern climates. The creatures could eat about seven hundred pounds of vegetation daily and lived sixty to eighty years. Mammoths are not the direct ancestors of modern elephants.

The Landmark is diligent in honoring the Native American heritage of the area. Ray Olachia, who was of Mescalero Apache heritage, and Bill Skillman, PhD, of Muscogee Creek heritage, were longtime volunteers at the Landmark. Both men have passed away, but they were active in helping with children's education projects. They also performed a ritual each fall to bless the ground of Yellowhouse Draw.

Skillman also discovered a non-native type of iris near the Singer store. An informational sign says that the yellow bearded iris grows in a circular pattern near the store's site, suggesting that it has been in that spot, undisturbed, for many years. The flower is believed to be a hybrid of two varieties cultivated initially in Nebraska. How it came to grow at the store is a mystery. The plant has been relocated to the back flower bed of the Interpretive Center.

Whether it is the bronzes, the trails, the Nash Interpretive Center exhibitions, the archaeological excavations, or the prairie itself, the Lubbock Lake Landmark has many subtle public educational aspects. The Landmark's archaeology is paramount; however, Johnson believes education is vital to its long-term preservation and stewardship. She and her team want people both to enjoy the Landmark's facilities and ecological beauties and to learn about the site's natural and archaeological history, following a tenet she learned from her mentor, Curry Holden.

CHAPTER 8

LESSONS FROM THE LANDMARK

Education has been vital to the Lubbock Lake Landmark since Curry Holden first identified the ancient Folsom points and bison bones found there in 1936. Holden knew that exploring and preserving what he suspected was a significant archaeological site would involve making the general public, city officials, and state lawmakers appreciate its importance. "Curry understood the need to both protect the site from further destruction while exploring its deposits and taking its 'secrets' to the public," Johnson writes in a chapter of the 2016 book *Interpreting First Peoples Bison Kills at Heritage Parks*.[1]

She calls Holden a visionary and counts him as a mentor. Johnson understands that people come to the Landmark for two reasons: science and the land. She knows that one group is interested in archaeology and the site's history, while others want only to spend time in the natural world. Holden's philosophy of public education and outreach programs is alive and well at the Landmark today. "Curry understood people's relationship to the landscape and how that experience shapes society and culture," Johnson says.

Johnson and her team believe archaeology is a powerful tool for today's humans to connect with the animals and the people who came before us. But their mission doesn't end with discovery. They are dedicated to interpreting the findings, understanding their significance, and devising innovative ways to share this knowledge with the general public. Whether through tours, trail signs, exhibitions, or community programs, Johnson and her team are committed to a broad-based effort to explain the last 2.6 million years and engage the public in understanding the importance of the past for the present and the future.

Holden began his campaign to preserve Lubbock Lake almost immediately by talking to anyone he could interest in the subject and taking groups on tours. He aimed to educate as many people as possible about the importance of Lubbock Lake as a Folsom site. He also brought Texas Tech students studying history and anthropology, subjects he taught, to the site. While tours continued irregularly over the next few decades, Johnson's arrival began a program of regular tours, particularly during the summer dig season. In 1973, her first summer at the Landmark, Johnson led the tours herself, delegating them to field crew members trained to lead tours the following summer and, ultimately, in the mid-1980s, to staff and volunteers who did not participate in excavations.[2]

Just as volunteers are the lifeblood of the excavations, so too are they essential to the programs offered by the Landmark. Today's volunteer corps comprises members of the South Plains Chapter of the Texas Master Naturalist Program and community members. Volunteers, also called docents, are not required to have any particular background; they only need an interest in the Landmark and the ability to engage with children and adults. Docents' roles vary based on their preferences. Their duties include leading tours, helping with children's activities, working at the front desk, helping

Figure 39. Curry Holden began leading tours almost immediately after identifying the area as an archaeological treasure. (Courtesy of Lubbock Lake Landmark)

visitors find their way around the site, answering phones, and working in the Landmark store.

Volunteering begins with a conversation with the Landmark's heritage education program manager, Susan Rowe. "I want to find out what interests each person [has], and then I know how to match them with the appropriate activity," she explains. There have been occasions when she discovered that the people asking to volunteer did not understand the Landmark. "I had one person who just wanted to tell people about dinosaurs," she smiles. "I had to tell him we do not have dinosaurs here. He came out to the Landmark several times, but he never volunteered."

All volunteers not working on excavations or in the Quaternary Research Center (QRC) go through a training program conducted by Rowe. Excavation or QRC volunteers receive training specific to their areas. Rowe's training includes talks by Johnson, heritage maintenance supervisor Scott Trevey, and Dr. Stance Hurst, plus a general overview of safety and Landmark operations. The training also takes the group into the galleries in the Nash Interpretive Center so they may acquaint themselves with the stories told there. That is when each volunteer begins to formulate how to give a tour.

Rowe prefers the terms "interpreter" and "facilitator" when giving her tour guides a title. She explains that docents function as interpreters who share information and explain the Landmark while guiding visitors through the galleries and as facilitators by providing visitors with a chance to connect with and understand the Landmark based on their interests. "We don't have prepared scripts; we have conversations," she explains. "Just as our facilitators will give tours differently based on their interests and expertise, visitors will respond to different areas of a tour based on their interests."

Richard Zartman, PhD, is a prime example. He became a Landmark docent in 2015 after retiring as chair of Texas Tech's Department of Plant and Soil Science. "I got tired of being a professional grandfather and decided I needed something else to do," he laughs. While his science background informs his tours, he mainly focuses on the peoples and cultures that have come through the area. "I want to talk about the cultures and make some of those animal bones come alive," he says.

One of Zartman's favorite topics is the ancient bison. He shows tour groups the bison statue outside the Nash Interpretive Center, asking them to imagine trying to kill a beast taller and much larger than they are with a flint tip attached to a piece of wood. From there, he explains how subsequent cultures developed new weapons to allow them to tackle the bison from further away, making hunting safer and more efficient. "Richard is great," says Rowe. "He will sit at the front desk or give tours. He is great at explaining Native American lifeways and is involved during our annual Bison Week, talking about ancient peoples and their tools and weapons."

Figure 40. During the annual Archaeology in Action week, Stance Hurst, left, and Doug Cunningham demonstrate the art of flintknapping, the creation of ancient stone tools. (Courtesy of Lubbock Lake Landmark; Kippra D. Hopper, photographer)

Even though they are given resources that explain the Landmark's history and the exhibitions, Rowe emphasizes that it is okay that facilitators do not have the answers to every visitor's question. "We encourage them to say, 'I don't know; let's see if we can find the answer,'" she explains. That may take the form of a closer examination of the exhibit interpretive panels or finding a staff member who can respond with more comprehensive information; sometimes a tour member with expertise in the subject can offer perspective. Rowe does not mind visitors expressing opinions or answering questions during a tour but is wary of the person who tries to take over the conversation and lecture rather than comment. "We've had that happen, including people who do not always believe what the science says and want to tell the group their personal opinions," she says. "We have to move the discussion back on topic tactfully."

Rowe's education and outreach efforts go beyond tours. She often speaks to local civic groups such as the Rotary and Lions

Clubs. She accepts invitations to set up booths about the Landmark at various events, including the Lubbock Arts Festival and the South Plains Fair. But the heart of her efforts is family-friendly activities. Rowe often hosts special days or weeks at the Landmark around a theme, such as Bison Week, which can draw more than a thousand people. An advantage of the Landmark's public programming is that every event is free.

Rowe also manages various free activities and summer youth programs for children ages three to twelve. During spring and summer, Growing Up Wild Saturdays is a two-hour program for three- and four-year-olds that includes stories, hands-on learning, and outdoor activities. Six weeks of summer programs for ages five to ten, with differing topics each week, fill up quickly. The program topics can range from studying bugs to migration patterns, fire ecology, art and nature, and archaeology and conservation—where participants process excavated materials and learn about field documentation and how an excavation works. Children can attend any or all of the week-long sessions.

The sessions are held mostly outdoors during the mornings, and Rowe makes it a point to offer a different experience. "We want to give them something they cannot get anywhere else," she says. "We may give them a worksheet, but then they have to go out and identify what is on it. How many bugs, birds, or flowers can you find?"

She and her volunteers encourage students to draw what they see or write about what they are learning. Rowe's goal is to give children a fun adventure that will also instruct them about the nature they see daily. "I'm going to do my best to teach about the world around them," she says. "They know more about the animals in Africa from watching television than they do about the animals here. I ask them if they know a fox can be in their backyard. Most do not. I want them to know about the animals, bugs, and birds here."

Figure 41. The Landmark's Summer Youth Programs quickly reach capacity. Session themes change to offer a variety of activities, such as basket flower seed harvesting. (Courtesy of Lubbock Lake Landmark)

Rowe's Conservation Academy is geared toward children eleven to twelve years old and offers a more in-depth, hands-on experience. The Academy features two two-week programs: the Agents of Deterioration course, led by Texas Tech's Heritage and Museum Sciences (HMS) master's degree students, and an advanced course, Digging Deeper, which is led by Landmark research staff and gives participants experience in the Quaternary Research Center. Sally Y. Shelton, associate professor of practice and associate chair of HMS, works with the Agents of Deterioration course. "The kids love to work with objects and find out what a museum does," Shelton says. Graduate students are in charge of Conservation Academy educational activities. Shelton marvels at Rowe's ability to interest students in heritage education. "Susan has attracted many of them to education when they never knew they would be interested in it and wound up changing their career paths."

Some camp participants are disappointed to find they will not learn about dinosaurs. Rowe says the staff explains that there were no dinosaurs at the Landmark. "I tell them that the only thing I know about dinosaurs is what I learned from *The Flintstones*," she laughs. "Sometimes, I have to explain the old television cartoon to them." However, she does not shy away from a conversation about dinosaurs and, during one camp session, explained that research now shows that some dinosaurs are the ancestors of birds. "One boy raised his hand and said, 'This weekend, we're going to hunt dinosaurs.' I asked what he meant, and he said his family was going duck hunting. I would love to have heard the conversation between him and his dad later," she says, smiling.

The boy may not have gotten it exactly right, but he made the connection and learned something, and that is the bottom line for Rowe. She is also grateful to the excavation staff who take the time to talk to the participants and explain how to identify a bison bone or what radiocarbon dating is. "Those kinds of things are not exactly my skill set, so it's wonderful when the real experts come and talk to the kids," she says.

The youth programs and Academy make a point of exposing the children to the excavations, visiting the dig sites, and seeing how what they are learning in camp is practically applied on a dig. Rowe is gratified when parents tell her how much their children talk about the camps. She remembers one young boy in particular. "He was so quiet, he barely talked at all. I was surprised when his mom told me that when he got in the car every day on the way home, he wouldn't stop talking about how much fun he had and what he had learned."

Another point of pride for Rowe is that the youth programs and Academy draw many repeat attendees. "We have had kids who have come together for several years. We had one pair of boys who met here every summer for some time," she smiles. "They were sad

Figure 42. The Landmark's Summer Youth Programs are designed for children ages three to twelve. Participants learn about the excavations at the Landmark by conducting their own digs. (Courtesy of Lubbock Lake Landmark)

when they reached the cutoff age. We do not have any programs for teens." However, once students reach age thirteen, they can participate in the Summer Youth Program as volunteers in the QRC laboratory. At age fifteen, they can volunteer for field crews.

José Mendoza and Keller Pettit are prime examples of two summer program participants who returned as volunteers. José started attending the youth programs at the Landmark when he was four and returned as a volunteer at thirteen. "The experience is fun!" says the outgoing, enthusiastic teen. "I love doing the activities. You get to learn about archaeology and so many other things."

Rowe changes the session topics each year to keep the program interesting. Although José particularly likes bugs, one of the sessions he loved best was the one on roping. "[Rowe] wanted us to lasso a fake cow," he says. "I had practiced roping before, so I do not want to brag, but I thought I would be good at it. Another kid had lassoed real cattle, so we had a competition." José reports that while he won some of the tosses, the other guy won more. But it is still a fond memory.

Now that he is a volunteer it is a different experience, but José still loves the program and plans to keep returning. He is an aide, helping Rowe and other staff keep track of the children. "It is an experience," he sighs. "I've worked with the youngest of the campers and some of the older kids. Both groups were interesting."

Besides the fun experience, José has many reasons for volunteering. It's how he likes to spend his free time, and he's partial to dragonflies and loves looking at their colorful wings.

"I love history, especially Native American history," he says. "My dad is part Apache, so I enjoy learning about his history. Miss Susan is so full of fun."

José's five siblings have all attended the Landmark youth programs at some time or other. His mother, Graciela Mendoza, appreciates the experiences the camps have given her children. "We homeschool, and I believe that these programs help them feel like they are in school," she explains. "They are not just out there looking around. They are learning."

Keller Pettit is José's opposite in demeanor, quiet and low-key. But his love of the Landmark youth programs matches José's. Keller recalls that he was about six years old when he first went to a program and returned as a teen volunteer until 2022, when he turned twenty and no longer had time.

Keller explains that his dedication to the programs comes from his love of history, even at age six. "The past has always been a big interest to me," he says. "The lives of people here go back 13,000 years; it is just so interesting." His favorite part of the Landmark? The dioramas in the Nash Interpretive Center. These displays depict early humans in natural surroundings with a painted mural background.

Keller relates that, as a volunteer, "I just wanted to ensure that the kids were feeling the same way I did about everything. There is so much out there, and I wanted them to have the same experiences I did." While he hasn't volunteered in a few years, in 2024, when we talked for this book, he had not ruled out becoming a volunteer again, but this time maybe in a different capacity, perhaps working on an excavation. "I think about it now and again," he says. "I think the excavations are oddly peaceful, and there is a careful order to them that makes sense to me."

Both José and Keller believe people should visit the Landmark for the programs, to see the exhibits, or just to walk around. "You don't have to spend a lot of time," Keller emphasizes. It is more than worth it."

The Landmark's education efforts go far beyond youth programs and tours. It is fertile ground for university students looking for experience in the heritage field or those trying to figure out precisely what they want to do as a profession. For Lucchese Hamilton, working at the Landmark opened up new options. "I was student teaching and looking for other education-related jobs," she recalls. "I did not love being in the classroom but still wanted to be in education." She found her answer during an internship at the Landmark in the Heritage Education division.

Hamilton began her internship in 2019 and joined the Texas Tech Heritage and Museum Sciences program the following year. She graduated in 2023 and landed her first job as the museum programs coordinator at the Bullock Texas State History Museum in Austin. The Texas Tech Master of Arts program, while housed in the museum, is overseen separately and reports to the dean of the Graduate School. The interdisciplinary program focuses on students working in the museum, at the Landmark, or in the QRC in collections processing, allowing them to develop a program that fits their long-term career goals.

For Hamilton, this opportunity opened new doors. Knowing she didn't want to become a classroom teacher, she concluded that a museum or heritage educator position at a historical or archaeological site was a good fit for her after her internship. The graduate program's experience-heavy nature gave Hamilton significant hands-on experience. "I didn't know anything about how museums work before I started here," she says. "This is absolutely the best place to be if you want to work in museums because of the direct experience you get."

Sally Shelton, the associate chair of HMS, agrees with Hamilton. She graduated from the master's program in 1984 and believes it gave her the foundation for a career that earned her a lifetime achievement award from the Registrar's Committee of the American Alliance of Museums in 2014. Her career has spanned several museums, including the National Museum of Natural History at the Smithsonian, where she was the collections officer; the San Diego Natural History Museum, as director of collections care and conservation; and the Museum of Geology at the South Dakota School of Mines and Technology, working with the construction of a new facility to organize conservation of the collections and write collection policies.

During the controversial and high-profile court case over who owned the skeleton of Sue, a *Tyrannosaurus rex* discovered in South Dakota in 1990, Shelton was a witness in a civil finding of fact hearing. Sue, named after the woman who first discovered the bones, is the largest and the most complete T-Rex skeleton ever found, standing thirteen feet tall and more than forty feet long. The question of who owned Sue was controversial and resulted in criminal and civil legal actions. The commercial company that excavated the skeleton claimed ownership, as did the landowner, who lived on the South Dakota Cheyenne River Sioux reservation. The company wanted to sell the skeleton. At the time, the federal government held the landowner's property in trust for him and seized the skeleton, declaring that permission to sell it had not been granted.[3]

The skeleton was coincidentally stored at the South Dakota School of Mines and Technology. When the company argued that the skeleton was theirs and could be damaged in storage, the government hired Shelton to assess its condition. Shelton issued a conservation report asserting that the skeleton was correctly stored and not in danger of damage. She did not go to work at the university for more than a decade after the hearing.

"It was a case of what you learned in the classroom turning up in real life," she says. "It was a difficult hearing. I was on the stand for almost two days. But thanks to my coursework, I knew how to write a conservation report. My time in the graduate program was long ago, but I am sure I learned how to write a report in Dr. Johnson's class." After years of civil and criminal litigation, the landowner was awarded the skeleton, which he sold at an auction to the Field Museum in Chicago in 1997.

While that was a memorable use of her graduate education, Shelton credits the HMS program for her overall career success. "Getting into the hands-on part of the program right away was important," she recalls. "It was not something you had to earn or where you had to start at a very low level. At Texas Tech, you are working from day one. Dr. Johnson has always advertised it as the perfect blend of the academic and practical. It was great."

The program usually has thirty to forty students enrolled each year. Most, if not all, students can participate in twelve-month paid apprenticeships that require them to work fifteen hours a week in the museum or at the Landmark. In addition to her duties as Landmark director, Johnson chairs the department and begins matching students with their curatorial area of interest during new student orientation. "[Students] work for a year in that area and can switch areas, if they want, for the second year," says Shelton. "Besides the paid internships, we have some scholarships and fellowships available. We try to provide as much financial support as we can."

The master's program, one of the oldest in the country, is typically two to three years long, concluding with a year-long paid internship, often at another institution, or a thesis project. There are two tracks: heritage science and museum science. It is one of the few known programs based entirely in a museum. Shelton, who returned to Texas Tech in 2020, notes that in her research for a report to the Texas Tech administration, she found only one program that is wholly museum-based: the Texas Tech degree program. "I went through almost every program out there, and most are based in a separate department that uses a museum as a resource or are largely online," she explains.

Hamilton is not the first graduate student to begin work at the Landmark before enrolling in the program. Lila Jones came to the Landmark as a volunteer in 2010 upon the recommendation of an undergraduate professor at the University of Costa Rica. The professor was a graduate of the HMS program. Rather than enrolling in the Heritage and Museum Sciences program, Jones enrolled in the interdisciplinary master's program at Texas Tech. Still, she worked at the Landmark and, in 2024, was research associate and senior crew chief at the Landmark and other regional dig sites. Madison Westfall is another master's program graduate who continued her career at the Landmark. She is the associate collections manager for the Quaternary Research Center.

Vance Holliday moved from volunteer to student to research collaborator with Johnson. He may well have been the first to take that route. "I started as an architecture major in junior college in San Antonio, but I quickly discovered I had no sense of design," he laughs. That's when he transferred to the University of Texas at Austin and majored in anthropology. He worked on contracts for different agencies in the archaeology office and thought it was a foot in the door.

After his graduation in December 1972, several things drew him to look at opportunities at Lubbock Lake. He had heard of the site during his undergraduate work and knew the team there would soon start a new excavation and research project. Holliday signed up to volunteer on a Landmark field crew, and after finishing a project he was already involved with, he arrived in Lubbock in the middle of the first field season in 1973. "What attracted me was that this was research, not just dealing with contracts," he says. "It was a fantastic experience, and it was interdisciplinary, involving geology, zooarchaeology, just everything I was interested in."

Holliday says he was already considering graduate school and knew the Texas Tech museum sciences graduate program was starting in the fall of 1974. "Eileen invited me to apply and told me I could work as a research assistant at the Landmark," he recalls. "It was just perfect for me." He returned to Austin briefly after the 1973 field season and came back to Lubbock in 1974 to enroll in the fall semester of the graduate program and continue his work at the Landmark. Holliday had the opportunity to work closely with the geology part of the excavations. He was particularly interested in soils and began taking classes in Texas Tech's Plant and Soil Science Department. He earned his master's degree and stayed on at the Landmark after graduation until 1978.

Holliday went on to earn a doctorate in geology at the University of Colorado in 1982. His dissertation was on the soils at the Landmark. From there, he taught and researched at Texas A&M and the University of Wisconsin–Madison. "I would always return and work at the Landmark when I could until I joined the University of Arizona," he says.

During his career, Holliday developed two main research interests: geology as it applies to answering archaeological questions, and Paleoindian archaeology, the study of early humans, which was a focus of the work at the Landmark. In 2002, he accepted a position at the University of Arizona that allowed him to combine these two research areas, focusing on the southwestern region of the United States. He retired in 2022 but continues to write and research on the early peoples of the Southwest. However, he has never lost interest in the Plains and maintains his relationship with Johnson and the Landmark.

Once the partnership with TPWD dissolved in 2000, the Landmark remained under the umbrella of the museum but would have a separate director, heritage education division, and staff. In addition to its anthropology holdings, generated primarily by the Landmark and its regional excavations, the museum holds significant collections in art, clothing and textiles, history, paleontology, and natural history, reflecting Curry Holden and the original Museum Association's desire to preserve the history of the area in all its forms. The museum holds more than nine million objects in its collections.

The museum's accreditation by the American Alliance of Museums (AAM) includes the Landmark. This designation is held by only a small percentage of all museums. The AAM's accreditation program's website states, "For more than 50 years, the Accreditation Program has been recognized as the gold standard of museum excellence."[4] Johnson led the team that completed the requirements to renew the museum's ten-year accreditation in 2020. She is also the museum's Senior Curator of Anthropology.

In 2024, the Institute of Museum and Library Services in Washington, DC, reported that there are more than 35,000 museums in the United States. The agency, which is the primary source for federal museum and library funding, includes a wide range of museums among its numbers, including botanical gardens, historical societies, zoos, and aquariums.[5]

The combination of academic and practical experience, as well as its wide-ranging collections and expertise in almost any field that might interest a student, makes HMS one of the top graduate museum programs in the country. Shelton notes that the academics combined with the practical experience give most of their students who graduate the credentials to secure a job immediately. "In 2024, statistics showed that about 92 percent of our students who graduate get a job at a museum or heritage site as their first job," she says.

In addition to her administrative and curatorial duties, Johnson teaches university classes specific to the museum grad program. Students take her collections management course early in the program, sometimes as a first course, and receive an introduction to the rigorous academic component. For Hamilton, it was the first course she took. "It was like diving in at the deep end," she remembers. "Dr. Johnson is tough, and everything she says in her lectures is essential. You must keep up with all the information she throws at you." Hamilton did fine in the course and became Johnson's teaching assistant the following year before interning in the Heritage Education Division.

Andrew DeJesse is another Heritage and Museum Sciences graduate student who was surprised by the difficulty of Johnson's course. He came to the program via a circuitous route. DeJesse is a colonel in the US Army Reserves. While he was deployed in Iraq and Afghanistan, his commander set him on a path combining his military career and his love of art; he was a graphic design major in college. DeJesse's new assignment as the Collective Heritage Lab director in the Army's Cultural Heritage Preservation Office was protecting art and heritage sites during wartime. When interviewed for this book in 2023, he was the Monuments Men and Women unit director.

Officially known as the Monuments, Fine Arts, and Archives program during World War II, the group had virtually disbanded between its height during the war and today. It drew attention from a nonmilitary audience after the 2014 George Clooney–directed movie *The Monuments Men*. The military created the unit to safeguard culturally significant works of art from destruction, working with Allied troops to avoid destroying significant cultural sites and assist with recovering works of art seized during the war.[6]

The mission remains virtually the same today. DeJesse advises senior military leadership on the responsibilities of the 1954 Hague Convention, which is designed to protect cultural sites and objects from the effects of armed conflict.[7] He is also charged with developing cultural heritage protection capabilities and recruiting cultural heritage experts into the Army. "We are not just art and museum curators," he emphasizes. "We are soldiers assigned to military units. We work with local museum professionals to discover what they believe is essential to their country and the community. We can also advise US and allied commands on 'no strike lists' to ensure the least damage to significant cultural sites."

While working to preserve cultural heritage sites in war zones, DeJesse realized he lacked the background to help cultural institutions fully. "I could walk into a museum and know how to keep the lights on and what security was required, but I had no clue about the museum side of how to care for collections, manage the objects, and develop policies," he says.

That is when he started looking at schools to broaden his knowledge base. Someone recommended the Texas Tech program to him. He and his family are based in Amarillo. His wife's family also lives there; many of them went to Texas Tech. After visiting with people in the program, DeJesse found exactly what he was looking for. "I know I was an unconventional student, but they left the door open for me to fill the gaps in my knowledge," he says. "Receiving my master's degree enables me to understand museums' needs and be better able to help our people be more successful."

As with Hamilton, one of the first courses DeJesse took was Johnson's collections management class, an area where he needed more knowledge. The class explains museum collections, general museum concepts, and issues related to managing and caring for collections. He also thought the class might be his last. "I took her first test, and I called my wife and said, 'I am done. I have just failed my test,'" he recalls, shaking his head. "But I squeaked by on the first test and figured out how to study for her tests. But it was traumatic."

He completed his degree and now teaches what Texas Tech calls a Maymester class in provenance theory and practice in the

Heritage and Museum Sciences program. A Maymester option is an intensive two-week course in which students can earn three credit hours. Provenance is the origin and history of an object and aids in restoring artifacts to their rightful owners and countries. "It's an honor to be a faculty member," he says. "You want to stay connected to things that are of quality. Texas Tech is different, and I feel a responsibility to tell the program's story as best I can through my experiences."

While Johnson's class may be the toughest in the curriculum, most students respect and remember her fondly. Sally Shelton can attest to this fact. Since joining the program's faculty, Shelton has facilitated annual alumni reunions at the Mountain Plains Museum Association meetings. The association covers a ten-state region offering its members programs, networking, and job information. "Eileen is the one person they always ask about. They almost all remark that they were terrified of her and were sure they bombed out of the program after the first test," Shelton laughs. "But we all survived, and in the end, you realize that she may be tough, but she is fair. She has never lowered her standards or allowed the program to lower its standards." For her part, Johnson admits to being tough but shakes her head and shrugs when she is told some people are afraid of her. "I really don't know why," she says.

Her students may not realize that Johnson had no intention of teaching a class when she first came to the Landmark. The program opened in 1974, and Johnson recounts that her sole focus was research and curation. She had always wanted to manage collections and was not interested in teaching. However, she did end up in the classroom, teaching her first class, a summer seminar, in 1976 and joining the faculty full-time in 1981 as a tenure-track assistant professor. "I have taught several courses over the years but have always taught collections management. I am still teaching collections management," she says, rolling her eyes. Johnson went on to receive tenure, and in 2007, Texas Tech named her a

Horn Professor, the highest honor accorded to a Texas Tech faculty member, recognizing their teaching and research.[8]

Johnson encourages professors and researchers in the sciences, arts, and humanities across the campus to utilize the Landmark, even if it is not always for an academic endeavor. The Texas Tech cross-country team and several high school and university runners train on the rolling trails. The Landmark is also a training ground for the South Plains Chapter of the Texas Master Naturalists, which holds classes at the facility.

While he is now retired, Richard Zartman is one of the faculty members who continues a small research project—one that had been started by a graduate student but was discontinued in the early 1990s—measuring groundwater fluctuations at the Landmark. If bones remain dry above the water table, they are safe from dissolving in water. Alternately, if there is standing water around the bones, they are saturated and not in danger of wearing away. When the water level changes, preservation issues can arise, leaving the bones covered with water at some points and dry at others. "If we have fluctuating water tables, then the bones could degrade over time," he explains.

Professors Susan Tomlinson and Jerod Foster draw on nature for their courses and use the Landmark as a classroom. In separate interviews, they shared many of the same thoughts about the Landmark's importance for students and society. When interviewed for this book, Associate Professor Tomlinson taught an interdisciplinary studies class in nature and the arts as part of Texas Tech's Honors College. She combines her undergraduate degree in art with her master's degree and doctorate in geology to teach students how to understand nature, write about it, and draw what they see.

When Tomlinson began teaching the class, she could not find an applicable textbook, so she wrote her own. The book is *How to Keep a Naturalist's Notebook*. In the preface, Tomlinson writes, "It is part lesson in the basics of sketching, part instruction in journal writing, and part walk through the process of identifying something and learning about it." She emphasizes that the book is not meant to teach students how to collect data for scientific study.

While the book was written for her class, Tomlinson notes that it helps amateur birdwatchers and professional scientists who want to learn how to maintain a nature notebook. The book offers tips on what one needs in the field and how to work there; it teaches basic drawing skills and how to identify birds and plants. It also gives examples of keeping field notes and offers examples of drawings and maps.[9]

The Landmark offers the perfect laboratory for Tomlinson's students. In addition to being a nature preserve, the site demonstrates how nature changes over the years. She explains why restoring the prairie was essential and talks about the unintended consequences of unchecked mesquite trees and the planting of elm trees in the 1940s. "We talk about reclaiming the land. How far back do you want to take it? One hundred years? Two thousand years?" she asks. "I tell students that this little patch of earth is as close as you can get to what it looked like in the 1880s."

Another lesson for Tomlinson's students is to think about the cycle of ecosystems. "As a geologist, I think in terms of millions of years. We talk about how the environment has changed over the centuries, how human intervention has affected our world," she says. "I tell students that every major species has gone extinct throughout the Earth's history. We are now the major species, and they should think about that."

A by-product of her course is that students, especially those raised in cities, get a new appreciation of nature. People who grew up in different parts of the country have different expectations of what a landscape should look like. "People think it is ugly here [on the Llano Estacado]," she explains. "They say the landscape is dead. And that is their first lesson: It is not dead. We talk about the resilience of the prairie plants and how they can survive drought and still flourish when they get rain."

Foster, an associate professor of practice at the Texas Tech College of Media & Communication, finds the Landmark the perfect training ground for his Adventure Media class. Foster is a professional photographer who has worked with *The New York Times*, The Nature Conservancy, *Texas Highways* and *Texas Parks & Wildlife* magazines, and the Bullock Texas State History Museum, among others. He focuses on stories that show the relationship between humans and nature.

Foster designed his Adventure Media class for students who wish to work in outdoor recreation media, conservation storytelling, and documentary filmmaking. The central part of the class happens during the university's spring break week. It involves a niche form of outdoor recreation called bike-packing. The class often rides hundreds of miles, using photography and video to document the experience. The class has visited Texas's Big Bend State Park, New Mexico's Lincoln National Forest, and the Organ Mountains region. "The Landmark is perfect for us; it is where we have our first class of the semester," Foster says. "Not only can the students get experience in off-pavement cycling, but they also get a feel for why paying attention to the land, the plants, and the animals around them is important."

In addition to preparing his students for the spring break trek, the Landmark speaks to Foster's love of the natural world. However, he and Tomlinson agree that getting used to the Plains can take a while. Tomlinson, who admits it took her some time to adjust to the Plains from her native New Mexico, applauds Johnson's efforts to educate the community about the Landmark. "The way to preserve something is to allow people to interact with it and get to know it," she explains. "Once they see it and have a relationship with the land, they love it."

Foster agrees and says he has developed an affinity with the Plains. "I am fascinated with the Great Plains. However, I think that many people must live on the Plains for a while before appreciating it," he states. Both professors describe the Plains as featuring a more subtle beauty than that presented by the mountains, the ocean, or vibrant green areas.

A few years ago, Foster began studying the Great Plains in depth. In 2024, he completed work on a documentary he calls *The Great Plains Project*, which he envisions as showcasing the Plains, the people, and the region's natural beauty. He and his collaborators initially planned to cycle from Lubbock to Calgary, Alberta, Canada, primarily on dirt roads. The route now extends south to Mexico. "The documentary features the Landmark," he says. "We will tell people to stop and learn about the history of the area and enjoy the restored prairie."

Foster believes restoring the native prairie was a significant addition to the Landmark. "Human development has largely impacted the region's natural ecology," he says, shaking his head. "It is essential to hold on to places like this. Whether you are interested in archaeology or do not even care to hike or bicycle, this is a great place to have lunch while you watch the hawks soar overhead or look at the prairie for a little bit."

Foster, like Tomlinson, sees the Landmark as vital to ensuring that future generations appreciate nature. "I know it is harder and harder to illustrate the significance of a place like this to younger generations, especially those who grow up in cities," Foster says. "It is one of the reasons I like to bring students here. Even students who grew up in Lubbock do not know this place. But once they see it, they can realize how beautiful it is."

Whether it is a research project, a class, a camp, or a tour, Johnson has worked for more than half a century to carry out Holden's mission to tell as many people as possible about the Landmark. In the 2016 book *Interpreting First Peoples Bison Kills at Heritage Parks*, Johnson's chapter discusses the site's preservation and public interpretation; she notes that the Landmark is much more than an archaeological locale. The site holds complete cultural and natural history records dating back to the early Quaternary, 2.6 million years ago. Water and food availability, usually in the form of bison, have drawn people here for at least 13,000 years, likely even longer.[10] The prairie restoration is a perfect complement to the archaeology. Johnson notes that native wildlife and plants have difficulty flourishing in a city and now have a natural living space at Lubbock Lake.

Johnson believes everything done at the Landmark is the foundation for public interpretation. Curry Holden understood from the beginning the relationship between research, public interpretation, and preservation. Johnson says that she and those at the Landmark are building on the vision and foundation built by Holden and Bob Nash, a foundation constructed through thoughtful public education and agreement between state agencies and the community.

"Our efforts are broad-based and interdisciplinary in nature, focusing on the relationships between the cultural and natural worlds," she emphasizes. "The Landmark can be viewed as a tool to engage the public in a forum on the importance of the past for the present and the present for the future."

CHAPTER 9

PAST, PRESENT, AND FUTURE

History often shows how the past shapes the future. As the University of Illinois History Department webpage eloquently put it in 2024, "History is the laboratory of human experience."[1]

The Lubbock Lake Landmark offers a stunning laboratory that allows visitors to explore how humans lived and adapted to the changing landscapes of the Southern High Plains. While many men and women have contributed to the research, education, and outreach at the Landmark, the site is fortunate that it has had long-term champions to preserve and protect it.

Eileen Johnson is the first to tell you that all she has accomplished has occurred due to university, community, city, county, and state support. "I have not done this alone; I'm grateful for all the help I've had over the years," she says. While Johnson and the Landmark have had several champions over the decades, three stand out. Each of these men marshaled patronage for the

Figure 43. William Curry Holden. (Courtesy of the Southwest Collection/ Special Collections Library, Texas Tech University)

Landmark. They also impacted Lubbock and Texas Tech in many other ways.

WILLIAM CURRY HOLDEN

From the time Curry Holden recognized the Landmark's significance in telling the story of early humans on the Southern High Plains, he relied on his background as a teacher to rally support for the fledgling archaeological site. Holden taught Johnson that the university and the community are tied together. "Curry felt strongly that the two benefited from each other," she says. "Curry understood people's relationship to the landscape and how that experience shaped society and culture."

Holden is often described as a visionary. His legacy at Texas Tech reaches far beyond the Lubbock Lake Landmark. He arrived at Texas Tech in 1929 to teach history and become the founding director of the university museum, serving until his retirement in 1964. As director, Holden oversaw the museum's early development, including leading fundraising efforts to build the institution's first building. The West Texas Museum opened in 1937. The museum, now named the Museum at Texas Tech University, moved into a new facility in 1970. To honor Holden, the university's Board of Regents in 1972 renamed the original building Holden Hall. That building now houses classrooms. While the structure has undergone expansion and renovations over the decades, the museum's original entry rotunda remains and holds a bust of Holden and a fresco he commissioned depicting life on the South Plains and luminaries of the day, painted by Peter Hurd.

As a professor, Holden led students on archaeology field trips in the West Texas area, including the two students who found the Folsom point that eventually led to the creation of the Lubbock Lake Landmark. In addition to identifying the Lubbock Lake Landmark's significance, he led an expedition in 1934 to observe and document the Yaqui, the Indigenous people living in Mexico. He would return to visit the Yaqui people several times.[2]

Holden's vision for the university was much broader than his work with the museum and the Landmark. He was also instrumental in forming the Southwest Collection/Special Collections Library, the National Ranching Heritage Center, and the International Center for Arid and Semi-arid Land Studies, all at Texas Tech.

ROBERT A. NASH

Robert "Bob" Nash became personally interested in the Landmark in the 1960s and grew to be a vocal advocate, leading efforts spanning about three decades to fund and preserve it.

Nash always had a story to tell, usually about the city or the Llano Estacado's history. Jeff Klotzman, a retired broadcast journalist, grew up down the street from Nash and knew him well. "Few people, if any, were as much of an expert on the history of Lubbock and the surrounding area," Klotzman says. Nash was an avid amateur historian, and Klotzman believes the story of Singer's store sparked his interest in the Landmark. "The store's history was part of this lengthy typewritten, single-spaced presentation that was almost academic in nature that he would give to civic clubs and other groups," he says.

Klotzman eventually worked for Nash at KAMC television and said it was an honor. He tells the story of covering the news conference at the Landmark, where it was announced that the new interpretive center would be named for Nash. "Morris Wilkes called us in advance, and we had a second camera there to capture [Nash's] reaction," Klotzman remembers. "Bob was just floored. It was a complete surprise and meant a lot to him."

Nash was active in the Lubbock Chamber of Commerce and served one term on the City Council. Lubbock insurance agent Alan Henry served as mayor during Nash's time in office. "Bob was such a public figure; everyone knew him and his radio voice," Henry says. "However, when it came to the council, he was not as vocal as you might think until he needed to be. He was a consistent advocate for the Landmark."

Johnson emphasizes that she doesn't believe that Nash received the recognition he deserved for his contributions to the Landmark. "Bob set into motion everything that we have at the Landmark," she says. "His influence built on Curry's influence with city and university leaders. He pushed the city and the museum to preserve the site."

JOHN T. MONTFORD

Bob Nash introduced then-Texas State Senator John Montford to Johnson and the Landmark. Montford used his connections to help move the Landmark forward. Johnson always believed it

Figure 44. John Montford (left) and Bob Nash at the ribbon-cutting event for the Nash Interpretive Center. (Courtesy of the Southwest Collection/Special Collections Library, Texas Tech University)

was Nash who piqued Montford's interest, but in an interview in 2023, Montford noted that while Nash influenced him, Johnson's passion really hooked him. "I was astonished at what had been uncovered at the Landmark," he says. "She made me realize that the site was significant not just in Texas but internationally. This was an important asset for Lubbock and a significant research opportunity. So, I felt like I wanted to help."

He brought his considerable influence in the state legislature to bear and helped shape what the Landmark is today. Montford is still particularly attracted to the ancient animals living there. "I had this fixation on the short-faced bear," he smiles at the memory. "He was a mean critter. And then, there was the giant armadillo. People said it was as big as a Volkswagen, and while that was an exaggeration, it was pretty big."

After leaving the State Senate, Montford became the first chancellor of the Texas Tech University System, raising more than $511 million in five years. He was also instrumental in convincing the state legislature to transfer ownership of the Landmark to the university. Montford would move on to AT&T and General Motors, holding top leadership roles. He is now president and CEO of JTM Consulting, LLC, and co-author of *Board Games: Straight Talk for New Directors and Good Governance*.[3]

KEEPER OF THE FLAME

Despite her reluctance to take credit for the Landmark's success, Eileen Johnson is its most enduring hero. As of 2026, she has overseen the Lubbock Lake Landmark for fifty-four years. Her tenacity and efforts to work around every *no* she received over that time have established the Landmark, and she herself, as internationally significant.

Johnson is not just a researcher. She is a teacher. Educating the general public and her students is a cornerstone of her career. Johnson has influenced numerous students, often partnering with them on excavations. It is a testament to her that former students, including those who are now colleagues, continue to seek her expertise.

In addition to garnering the university's highest faculty award,[4] Johnson has received recognition from her peers. The journal

Quaternary International recognized her career by dedicating a 2018 issue to her and her research. In the opening editorial, journal editors wrote, in part, "We propose to honor the career of Eileen Johnson because we believe her to be a complete scientist. We use this term because she represents a complete scientific career, going beyond pragmatic training and theoretical issues to include public awareness and issues of ethics."[5]

In 2024, she received the Jack Nokes Outstanding Service Award from the Texas Association of Museums (TAM). The organization's website states that the award recognizes significant contributions to the museum field, service to TAM, and truly remarkable individuals. In 2016, she received the Hugo G. Rodeck Award for Excellence from the Mountain Plains Museum Association for her service to the organization and the museum profession. Vance Holliday, one of Johnson's oldest colleagues and friends, isn't surprised at the honors she has received. "Eileen is so smart," he says. "She thinks broadly and understands the breadth of archaeology. She pushed interdisciplinary archaeology, and that was unusual and remarkable."

Holliday calls Johnson persistent and a survivor. She admits she can be obstinate, a trait that helped her achieve early success. When Johnson entered the field, she was one of only a handful of women. She still shakes her head as she recounts being pregnant while working on excavations in the hot summer sun. "It wasn't easy; it's hard physical work, and on top of that, I was trying to get a neophyte research program off the ground," she says. "By the time I came along, I was lucky that there were a few women in the field and a few who were leading research projects. Honestly, I didn't think much about it. I was going to do what I was going to do. I was stubborn and persistent. If someone didn't want to work for a woman, then fine, don't come here."

Johnson notes that her original mentors, Craig Black and Curry Holden, did not have a problem taking women into the field. "Curry was an incredible trailblazer. He took women, including his daughter Jane, into the field." Johnson leans forward and taps her desk while talking about her early female colleagues. "Jane grew up in the field. She did her own fieldwork. There were a few others in the 1950s and 1960s, but you're talking only a handful."

When it's pointed out that she, too, is a trailblazer, she demurs, indicating her reluctance to make the Landmark about her. "I don't think I ever thought about it at the time," she says, pausing to choose her words carefully. "I didn't let it get to me, or I dealt with the issue by pushing forward, being persistent, doing research, and publishing. People who wanted to prevent me from doing the work, or wanted to take over, or take the credit, weren't doing research. They weren't publishing. I was."

While her early years were difficult, she notes that she was never really stymied; she worked to find another way. "I never set out to break barriers, and I didn't do those things consciously," she muses. "I think I did what I knew would further my research efforts." Today, she oversees Texas Tech's Heritage and Museum Sciences master's degree program, where the students are primarily female. While few of those students go into some form of archaeological work, Johnson notes that the archaeology field is becoming primarily female.

What does the future hold for the Landmark? There's no crystal ball to reveal that answer. However, one thread that recurred throughout the interviews for this book is that the preservation of the site must continue simply because of its significance. Holliday believes people have an intrinsic interest in the past. "People love dinosaurs and mammoths," he explains. "The Lubbock Lake Landmark is a great way to portray history for people, from mammoths to how the environment has changed, to the city's founding. It's all right here in their backyards."

Aaron Pan, PhD, executive director of the Museum at Texas Tech University, sees the Landmark as a phenomenal resource and credits Johnson's long-term leadership in establishing methodical and meticulous documentation of everything extracted from an excavation. "Due to her long tenure, you have a seamless sequential discovery, documentation, preservation, and conservation process," he explains. He also marvels at the number of objects the Landmark has contributed to the museum's collection. "The museum now sits at a little more than 9.5 million objects and specimens, about 45 percent from the anthropology collection," he declares.

Johnson is well aware that she won't be around forever and has hired people to continue the Landmark's growth as positions open up. In 2024, she hired a new operations manager and a research aide to inventory wildlife at the Landmark.

Joseph Pitts is the Landmark's operations manager. He has financial oversight of the Landmark's operations and gift shop budgets and is responsible for its general upkeep, including the trails, interpretive center, research area, and field camp. In addition to all the administrative duties, he describes himself as an "extra pair of hands" when an area needs assistance. "I try to help every area do what it needs to do," he explains. "I may help the education area if they need extra help with a public program or a tour. Or I'll pitch in and help Scott Trevey if he needs assistance maintaining the trails or grasslands."

A significant part of Pitts's role is preparing the Landmark excavation sites for the summer field season. That is a time-consuming effort that includes everything from ensuring that housing is ready for volunteers and other facilities are in good working order to cleaning the sites. After taking the job, one of the first things on his agenda was helping Trevey remove the old tents and install new housing for volunteers at the Landmark for the 2025 season.

A 2021 Texas Tech Heritage and Museum Sciences program graduate, Pitts says his goal is to raise public awareness of the site and its significance. "I want visitors to leave here feeling inspired, happy, and eager to share their experiences with others," he says.

Johnson hired Shea Mullican in 2024 to work on a three-year

Figure 45. Lubbock Lake Landmark at dusk. (Courtesy of Lubbock Lake Landmark; photographer, Kippra D. Hopper)

grant to inventory the animals that live at the Landmark. During her time pursuing her master's degree at Texas Tech, Mullican worked on a project examining kestrel reproduction habits. "The birds don't build their own nest, preferring to nest in existing holes or structures. One of the sites was at the Landmark, and I became familiar with the staff and the site," she says.

That project led to her position as a research aide. She is completing an inventory of the vertebrate species at the Landmark using cameras, personal observation, and anecdotal sightings from people who visit the site. It's not easy to spot the birds and critters who call the Landmark home. "Hopefully, when we're done, people in the future can look back and know what lived here today and repeat the study to see how the habitat and animals have changed," she says.

When Mullican looks to the future, her biggest wish is that urban sprawl not change the Landmark. "I have a perfect idea of what I want the Landmark to look like," she says. "However, I'm a realist, and I know it's possible that the city might encroach on the site. I want the wildlife to have a safe haven. I also want the public to be able to come here and see the wildlife going about their day-to-day lives."

As for Johnson, she sees her future as continuing at the Landmark. "I'm not ready to retire; there's too much still to do, too much that I want to do," she emphasizes. "I want to continue to research and be at the Landmark."

However, Johnson too is realistic. She knows that much of what remains buried at the Landmark will not be unearthed in her lifetime. Only five to six percent of the Landmark's 336 acres have been excavated. The water table is unlikely to recede enough to allow researchers to dig along the reservoir's banks and bottom, where the oldest objects may lie, at least not anytime soon. Johnson also knows that there are significant sites outside the Landmark's boundaries. Still, those areas are in private hands or owned by the city. She doubts that Landmark researchers will ever gain access to them.

"Thanks to our state and federal designations, the current Landmark should be preserved," she says. "I know I won't get to everything, and that's fine. It's in the bank for the next generations." So, if William Shakespeare was correct when he wrote in *The Tempest* in 1611 that the past is prologue, perhaps a new champion will emerge to continue Johnson's fight to preserve and continue the research at the Landmark.[6]

Ultimately, Lubbock Lake Landmark's future lies in the hands of the university leadership and ongoing efforts to secure funding for research and education. Those who love the Landmark for its rich history, scientific importance, and natural beauty hope that future university and museum leaders will remain committed to protecting what is recognized internationally as one of the most significant hunter-gatherer archaeological sites in the New World.

ACKNOWLEDGMENTS

I have many people to thank for making this book a reality. However, not a single word would have been possible without Eileen Johnson's help, patience, and encouragement. Although she follows in the footsteps of the exceptional W. Curry Holden—who first recognized the site's importance—and other outstanding scientists who came before her, make no mistake: the Landmark is not only her work but also her passion. By the time of publication, she will have spent more than fifty years nurturing the site, conducting research, and educating future museum and quaternary sciences professionals.

Although the writing process is solitary, I was fortunate to have a network of friends and family who provided unwavering support. I am particularly grateful to my sisters. Susann Follis is my sister by blood. She always offered encouragement and support and kept telling me to hurry up and finish the next chapter so she could read it. My chosen sisters, Lisa Low and Jo Nelon, were always willing to let me vent or discuss challenging issues and help me get to the finish line.

I also must thank Kippra Hopper, a longtime friend and employee at the Landmark, for her invaluable help with research and proofreading and for serving as a sounding board and cheerleader. Stance Hurst and John Moretti were essential sources of information about the excavations, animal bones, and human artifacts.

Many people, including Vance Holliday, who started with Johnson during her first field season, have gone on to distinguished careers. Professor Holliday is a valuable source of information, and he generously offered his time to fact-check parts of the book. He also spent a considerable amount of time trying to help me understand radiocarbon dating.

Scott Trevey, Doug Cunningham, Susan Rowe, and Lila Jones were also indispensable sources, and I thank them for taking the time they didn't really have to talk with and encourage me. The folks at the Landmark were so kind to make me part of their family while I researched and wrote the book.

Judith Keeling, a longtime friend and editor extraordinaire, played a significant role in my finishing this book. While I've made my living as a writer, writing a book is a new experience. Judith's editing skills and suggestions helped me figure out how to write the book, and I do not doubt that she made it better.

A big thank-you to Joanna Conrad, Travis Snyder, and everyone at Texas Tech University Press for their guidance, encouragement, and patience.

To Lynn Whitfield, Marshall West, and all the staff of the Southwest Collection/Special Collections Library, thank you for pulling box after box of materials and oral histories and patiently answering my endless questions.

The end result is more than one hundred hours of interviews, two file drawers full of paper, a couple of jump drives of material, and a trip to an active dig site to see the work firsthand.

I never imagined how these disparate pieces would weave into a story that must be told and retold. I am grateful to everyone who guided me.

NOTES

CHAPTER 1

1. "Quaternary Period," *National Geographic*, https://www.nationalgeographic.com/science/article/quaternary.
2. "Geology of Rocky Mountain National Park," Geology and Ecology of National Parks, United States Geological Survey, https://www.usgs.gov/geology-and-ecology-of-national-parks/geology-rocky-mountain-national-park.
3. Eileen Johnson, "Landscapes and Peoples of the Llano Estacado," in *Archaeological Landscapes on the High Plains*, ed. Laura L. Scheiber and Bonnie J. Clark (Denver: University Press of Colorado, 2008), 124.
4. Johnson, "Landscapes and Peoples of the Llano Estacado," 124.

CHAPTER 2

1. Eileen Johnson, "Landscapes and Peoples of the Llano Estacado," in *Archaeological Landscapes on the High Plains*, ed. Laura L. Scheiber and Bonnie J. Clark (Denver: University Press of Colorado, 2008), 131.
2. W. Curry Holden, "Historical Background of the Lubbock Lake Site," *The Museum Journal* 15 (1974): 13.
3. Blackwater Draw National Historic Landmark, Eastern New

Mexico University, 2017–2026, https://www.bwdarchaeology.com/bwdnationalhistoriclandmark.

4. Matt Peoples, "George McJunkin and the Discovery That Changed American Archaeology," *Archaeology Southwest* (blog), February 23, 2015, https://www.archaeologysouthwest.org/2015/02/23/george-mcjunkin-and-the-discovery-that-changed-american-archaeology/.
5. Joe Ben Wheat, "First Excavations at the Lubbock Lake Site," *The Museum Journal* 15 (1974): 15.
6. Eileen Johnson, "Lubbock Lake Landmark – A Joint Venture in Preservation, Research, and Public Interpretation," in *Interpreting First Peoples Bison Kills at Heritage Parks*, ed. Leslie B. Davis and John W. Fisher Jr. (Salt Lake City: University of Utah Press, 2016), 132.
7. Curry Holden, "The Land," *A History of Lubbock*, Part 1, edited by Lawrence L. Graves, West Texas Museum Association, 1959, 19.
8. Holden, "The Land," 20.
9. Johnson, "Lubbock Lake Landmark – A Joint Venture," 133.
10. Eileen Johnson and Vance T. Holliday, "Introduction," in *Lubbock Lake Late Quaternary Studies on the Southern High Plains*, ed. Eileen Johnson (College Station: Texas A&M University Press, 1987), 7.
11. Jane Holden Kelley, "A Brief Resume of Artifacts Collected at the Lubbock Lake Site Prior to 1961," *The Museum Journal* 15 (1974): 43.
12. Johnson, "Lubbock Lake Landmark – A Joint Venture," 133.
13. Johnson, "Lubbock Lake Landmark – A Joint Venture," 133.
14. Charles Mann, "The Clovis Point and the Discovery of America's First Culture," *Smithsonian*, November 2013, https://www.smithsonianmag.com/history/the-clovis-point-and-the-discovery-of-americas-first-culture-3825828/.
15. Vance T. Holliday, "Cultural Chronology of the Lubbock

Lake Site" (master's thesis, Texas Tech University, Department of Museum Science, 1977), 18.

16. Britannica editors, "Folsom Complex," *Encyclopedia Britannica*, January 5, 2012, https://www.britannica.com/topic/Folsom-complex.
17. C. Dean Wilson, "Chupadero Black-on-White," New Mexico Office of Archaeological Studies Pottery Typology Project, 2012, https://ceramics.nmarchaeology.org/typology/type?p=287.
18. Library of Congress, "The American West, 1865–1900," U.S. History Primary Source Timeline, accessed July 23, 2025, https://www.loc.gov/classroom-materials/united-states-history-primary-source-timeline/rise-of-industrial-america-1876-1900/american-west-1865-1900/.
19. *Galveston Daily News*, April 26, 1886.
20. "Monte Verde Archaeological Site," UNESCO World Heritage Convention, submitted February 23, 2004, https://whc.unesco.org/en/tentativelists/1873/.
21. Dave Roos, "How Early Humans First Reached the Americas: 3 Theories," *History*, last updated March 17, 2025, https://www.history.com/articles/human-migration-americas-beringia.
22. Robert Sanders, "Tests Confirm Humans Tramped Around North America More Than 20,000 Years Ago," *Berkeley News*, October 5, 2023, https://news.berkeley.edu/2023/10/05/tests-confirm-humans-tramped-around-north-america-more-than-20-000-years-ago.
23. Vance T. Holliday, et al., "Paleolake Geochronology Supports Last Glacial Maximum (LGM) Age for Human Tracks at White Sands, New Mexico," *Science Advances* 11, no. 25 (2023): eadh5006, https://www.science.org/doi/10.1126/sciadv.adv4951.
24. Roos, "How Early Humans First Reached the Americas."

25. Roos, "How Early Humans First Reached the Americas."
26. Shannon McIntire, "Can You Dig It?" Texas Department of Transportation Newsroom, May 13, 2025, https://www.txdot.gov/about/newsroom/stories/can-you-dig-it.html.

CHAPTER 3

1. Eileen Johnson, "Lubbock Lake Landmark – A Joint Venture in Preservation, Research, and Public Interpretation," in *Interpreting First Peoples Bison Kills at Heritage Parks*, ed. Leslie B. Davis and John W. Fisher Jr. (Salt Lake City: University of Utah Press, 2016), 129.
2. Johnson, "Lubbock Lake Landmark – A Joint Venture," 138.
3. Johnson, "Lubbock Lake Landmark – A Joint Venture," 134.
4. Johnson, "Lubbock Lake Landmark – A Joint Venture," 135.
5. Johnson, "Lubbock Lake Landmark – A Joint Venture," 138.
6. Johnson, "Lubbock Lake Landmark – A Joint Venture," 138.
7. Jerry McCarty, "Lack of Funds May Delay Lubbock Lake Site Park," *Lubbock Avalanche-Journal*, October 13, 1967.
8. Jerry McCarty, "Park Development Plans Aired Here," *Lubbock Avalanche-Journal*, April 5, 1968.
9. Johnson, "Lubbock Lake Landmark – A Joint Venture," 139.
10. Johnson, "Lubbock Lake Landmark – A Joint Venture," 139.
11. Johnson, "Lubbock Lake Landmark – A Joint Venture," 141.
12. Johnson, "Lubbock Lake Landmark – A Joint Venture," 143.
13. Jim Goss, oral history interview by Daniel U. Sánchez, online transcription, October 8, 2014, Southwest Collection/Special Collections Library, Texas Tech University, https://swco-ir.tdl.org/server/api/core/bitstreams/8b269ad9-001e-4abc-8854-c15657e0e9e7/content.
14. Johnson, "Lubbock Lake Landmark – A Joint Venture," 143.
15. Johnson, "Lubbock Lake Landmark – A Joint Venture," 143.
16. Johnson, "Lubbock Lake Landmark – A Joint Venture," 143.
17. Johnson, "Lubbock Lake Landmark – A Joint Venture," 144.

18. Johnson, "Lubbock Lake Landmark – A Joint Venture," 144.
19. Johnson, "Lubbock Lake Landmark – A Joint Venture," 144.
20. Johnson, "Lubbock Lake Landmark – A Joint Venture," 143.
21. John T. Montford and Joseph Daniel McCool, *Board Games: Straight Talk for New Directors and Good Governance* (n.p.: Praeger, June 30), 201, https://boardgamesbook.com/authors/john-t-montford/.
22. Johnson, "Lubbock Lake Landmark – A Joint Venture," 145.
23. Texas Parks and Wildlife, *Lubbock Lake Landmark State Historic Site, Master Plan and Site Analysis*, January 1988, 59.
24. Wendy E. Lane, "Experts: Lubbock May Be One of New World's Oldest Towns," *Dallas Morning News*, December 26, 1988.
25. Johnson, "Lubbock Lake Landmark – A Joint Venture," 150.
26. Johnson, "Lubbock Lake Landmark – A Joint Venture," 151.
27. Johnson, "Lubbock Lake Landmark – A Joint Venture," 152.
28. Johnson, "Lubbock Lake Landmark – A Joint Venture," 152.

CHAPTER 4

1. "About Archaeology," Society for American Archaeology, https://saa.org/Member/SAAMember/About-Archaeology/What-do-Archaeologists-do.aspx?hkey=7e6707b0-83ff-446f-9cc5-768220459a59.
2. "Cosmic Origins: Giving the Clock Its Tick," Lubbock Lake Landmark Interpretive Center, permanent exhibit, opened 1990.
3. "About Archaeology," Society for American Archaeology.
4. "About Archaeology," Society for American Archaeology.
5. "About Archaeology," Society for American Archaeology.

CHAPTER 5

1. William Curry Holden, "U Lazy S Ranch," Texas State Historical Association, updated January 23, 2019, https://www.

tshaonline.org/handbook/entries/u-lazy-s-ranch.

2. Holden, "U Lazy S Ranch."
3. "American Lions," National Park Service, January 27, 2002, https://www.nps.gov/whsa/learn/nature/american-lions.htm.
4. Eileen Johnson and John A. Moretti, "Diversity and Paleoecology of Zone 3 Late Pleistocene Vertebrates at Kincaid Shelter (41UV2), Central Texas, USA," *PaleoAmerica: A Journal of Early Human Migration and Dispersal* 8, no. 3 (2022): 228–52.
5. Stance Hurst and Eileen Johnson, "Prominent Places on the Landscape: Occupations at Cowhead Mesa Along the Eastern Escarpment Edge of the Southern High Plains of Texas," *Plains Anthropologist* 62, no. 241 (2016): 5–31, https://www.tandfonline.com/doi/fll/10.1080/00320447.2016.1235303.
6. Hurst and Johnson, "Prominent Places on the Landscape," 5–31.
7. Stance Hurst, Elizabeth Louden, and Eileen Johnson, "Unravelling the Cowhead Mesa Petroglyphs with 3-D Laser Scanning and High-Resolution Digital Photography," *Rock Art Research* 26, no. 1 (May 2009): 43–53, https://www.researchgate.net/publication/265085921_Unravelling_the_Cowhead_Mesa_Petroglyphs_with_3-D_Laser_Scanning_and_High-Resolution_Digital_Photography.
8. Kristina Woods Butler, "Preserving History for the Future," *Discoveries*, Spring 2012, Texas Tech University.
9. Stance Hurst, Dallas C. Ward, Eileen Johnson, and Doug Cunningham, "Cowboy Life Along the Llano Estacado Eastern Escarpment of Northwest Texas: Insights from Macy Locality 16," *Historical Archaeology* 52 (March 30, 2018): 332–47.
10. Alyson L. Greiner, "Dugouts," *Encyclopedia of Oklahoma History and Culture*, https://www.okhistory.org/publications/enc/entry?entry=DU003.

11. "Bison, Buffalo, Tatanka: Bovids of the Badlands," Wildlife in the Badlands, National Parks Service, https://www.nps.gov/articles/bison_badl.htm.
12. "Whose Buffalo? - Fact Sheet," Texas Beyond History, https://www.texasbeyondhistory.net/redriver/images/rr-whosebuffalo.pdf.
13. J. Brett Cruse, *Red River War: Archeological Perspectives on the Indian Campaign of 1874*, (Austin: Texas Historical Commission, 2017, 2nd printing), 1.
14. James L. Haley, *The Buffalo War: The History of the Red River Indian Uprising of 1874* (Austin: State House Press, 1998), viii.
15. J. W. Woody, letter to J. Evetts Haley, October 10, 1926, J. Evetts Haley Collection, Haley Memorial Library and History Center, Midland, Texas.
16. J.W. Woody, letter to J. Evetts Haley, February 11, 1928, J. Evetts Haley Collection, Haley Memorial Library and History Center, Midland, Texas.
17. Hugh Allen Anderson, "Black Horse [TuUkumah]," *Handbook of Texas Online*, Texas State Historical Association, accessed June 9, 2025, https://www.tshaonline.org/handbook/entries/black-horse-tu-ukumah.
18. Herman Lehmann, *Nine Years Among the Indians, 1870–1879* (n.p.: Arcadia Press, 2019), 85.
19. Wayne Gard, *The Great Buffalo Hunt* (Alfred A. Knopf, 1959), 238.
20. Hugh Allen Anderson, "Yellow House Canyon, Battle of," *Handbook of Texas Online*, Texas State Historical Association, accessed June 29, 2026, https://www.tshaonline.org/handbook/entries/yellow-house-canyon-battle-of.

CHAPTER 6

1. Vance T. Holliday, "Eileen Johnson's Legacy of Research," *Quaternary International* 492 (October 2018): 3, https://

www.sciencedirect.com/journal/quaternary-international/vol/492/suppl/C.

2. Stance Hurst and Eileen Johnson, "Gearing Up at the Adair-Steadman (41FS2)," *PaleoAmerica* 2, no. 3 (August 18, 2016): 252–60, https://www.tandfonline.com/doi/citedby/10.1080/20555563.2016.1198215?scroll=top&needAccess=true.
3. *Engaging Folsom (10,800–10,200) Hunter-Gatherers with 3D Technologies*, Lubbock Lake Landmark Nash Interpretive Center exhibit, 2017, https://www.depts.ttu.edu/museumttu/lll/EngagingFolsom.html.
4. *Engaging Folsom (10,800–10,200) Hunter-Gatherers with 3D Technologies*, Lubbock Lake Landmark.
5. Stance Hurst and Eileen Johnson, "Adair-Steadman (41FS2) – Survey at a Folsom Site in the Brazos River Drainage on the Southern Plains," *Index of Texas Archaeology* 1, no. 26 (January 2019), https://scholarworks.sfasu.edu/ita/vol2019/iss1/26/.
6. *Engaging Folsom (10,800–10,200) Hunter-Gatherers with 3D Technologies*, Lubbock Lake Landmark.
7. "Marcy, Randolph Barnes (1812–1887)," *Encyclopedia of Oklahoma History and Culture*, Oklahoma Historical Society, https://www.okhistory.org/publications/enc/entry?entry=MA021; Thomas W. Cutrer, "Marcy, Randolph Barnes (1812–1887)," *Handbook of Texas Online*, Texas State Historical Association, December 9, 2020, https://www.tshaonline.org/handbook/entries/marcy-randolph-barnes.
8. Taylor Wiley, Valeria Chamorro, Joshua J. Wells, Eric Kansa, and Patrick Finnegan, "More Than Just Another Number: Use of the Smithsonian Trinomial System and the Digital Index of North American Archaeology (DINAA) to Link Open Information about Archaeological Sites Across the Web," Society for American Archaeology 82nd Annual

Meeting, Vancouver, BC, March 2017.

9. Wiley, et al., "More Than Just Another Number."
10. American Museum of Natural History, *The Evolution of Horses*, American Museum of Natural History Part of the Horse exhibition, May 17, 2008–January 4, 2009, https://www.amnh.org/exhibitions/horse/the-evolution-of-horses.
11. John Moretti, "Pleistocene Proboscidean Tusk Segments Discovered," *Notes from the Field* (Lubbock Lake Landmark newsletter), Winter 2013, https://www.depts.ttu.edu/museumttu/lll/newsletters/Winter%202013.pdf.
12. Anna Jean Taylor, *Pastores in the Texas Panhandle, 1876–1884* (thesis, Wayland Baptist College, Spring 1976).
13. J. Kent Hicks and Eileen Johnson, "Pastores Presence on the Southern High Plains of Texas," *Historical Archaeology* 34, no. 4 (2000): 55, http://www.jstor.org/stable/25616851.
14. Hicks and Johnson, "Pastores Presence," 55.
15. Hicks and Johnson, "Pastores Presence," 47.
16. Hicks and Johnson, "Pastores Presence," 55.
17. "Quanah Parker," The Historical Marker Database, December 21, 2017, https://www.hmdb.org/m.asp?m=111756.

CHAPTER 7

1. Art Leatherwood, "Llano Estacado," *Handbook of Texas Online*, Texas State Historical Association, June 18, 2020, https://www.tshaonline.org/handbook/entries/llano-estacado,
2. Leatherwood, "Llano Estacado."
3. George J. Goodman and Cheryl A. Lawson, *Retracing Major Stephen H. Long's 1820 Expedition* (Norman: University of Oklahoma Press, 1995), xii.
4. John Cooper Jenkins, *Estacado: Cradle of Culture and Civilization on the Staked Plains of Texas* (Crosby County Pioneer Memorial Museum, 1986), 56.

5. Jenkins, *Estacado*, 56.
6. Jenkins, *Estacado*, 56.
7. Eileen Johnson, "Landscapes and Peoples of the Llano Estacado," in *Archaeological Landscapes on the High Plains*, ed. Laura L. Scheiber and Bonnie J. Clark (Denver: University Press of Colorado), 117.
8. Leatherwood, "Llano Estacado."
9. Leatherwood, "Llano Estacado."
10. Texan by Nature, "A Conservation Guide for Texas Landowners," https://texanbynature.org /2023/03/a-conservation-guide-for-texas-landowners/.
11. "King Ranch," https://king-ranch.com/about-us/maps/.
12. "Ethnobotany of Mesquite Trees," National Park Service, July 4, 2020, https://www.nps.gov/articles/000/ethnobotany-of-mesquite-trees.htm.
13. Eileen Johnson and Vance T. Holliday, "Introduction," in *Lubbock Lake: Late Quaternary Studies on the Southern High Plains*, ed. Eileen Johnson (Texas A&M University Press, 1987 second printing, 2007), 7.
14. Emily Osterloff, "Tumbleweeds: The Fastest Plant Invasion in the USA's History," Natural History Museum, London, accessed May 18, 2024, https://www.nhm.ac.uk/discover/tumbleweeds-fastest-plant-invasion-in-usa-history.html.
15. *Engaging Folsom (10,800–10,200) Hunter-Gatherers with 3D Technologies*, https://www.depts.ttu.edu/museumttu/lll/EngagingFolsom. Lubbock Lake Landmark Nash Interpretive Center exhibition, 2017–2018.
16. "Fire on the Land: Native People and Fire in the Northern Rockies," Confederated Salish and Kootenai Tribes – Division of Fish, Wildlife, Recreation, and Conservation, last modified 2021, http://fwrconline.csktnrd.org/Fire/FireOnTheLand/.
17. Geoffrey Donovan and Thomas C. Brown, "Be Careful What

You Wish For: The Legacy of Smokey Bear," *Frontiers in Ecology and the Environment* 5, no. 2 (2007): 73–79.

18. Donovan and Brown, "Be Careful What You Wish For," 73–79.
19. "Smokey Bear About Wildfires," Ad Council, https://smokeybear.com/.
20. "Henry A. Wright Memorial Endowed Scholarship," Davis College of Agricultural Sciences and Natural Resources, Texas Tech University, September 28, 2023, https://www.depts.ttu.edu/agriculturalsciences/Students/scholarships/nrm/wright.php.
21. "Prescribed Burning Board,"Texas Department of Agriculture, https://texasagriculture.gov/Home/Production-Agriculture/Prescribed-Burn-Program/Prescribed-Burning-Board.
22. Sarah Wade, "8 Surprising Prairie Dog Facts,"World Wildlife Fund. https://www.worldwildlife.org/resources/explainers/prairie-dogs-the-tiny-architects-of-the-great-plains/.
23. Wade, "8 Surprising Prairie Dog Facts."
24. Cullom Simpson, Whitney Gann, Bonnie Warnock, and Louis A. Harveson, Borderlands Research Institute, Sul Ross State University, May 27, 2021, https://bri.sulross.edu/project-spotlight-prairie-dogs-and-cattle/#.
25. Daria McKelvey, "The Beauty of the Lubbock Lake Landmark," *Notes from the Field* (Lubbock Lake Landmark newsletter), Summer 2020.
26. Vance T. Holliday, Susan D. Hovorka, and Thomas C. Gustavson, "Lithostratigraphy and Geochronology of Fills in Small Playa Basins on the Southern High Plains," *Geological Society of America Bulletin* 108 (1996): 953–65.
27. Vance T. Holliday, "Guidebook to the Central Llano Estacado," *Friends of the Pleistocene South-Central Cell Field Trip* (ICASALS and the Museum of Texas Tech University).
28. Eileen Johnson, "Lubbock Lake Landmark – A Joint Ven-

ture in Preservation, Research, and Public Interpretation," in *Interpreting First Peoples Bison Kills at Heritage Parks*, ed. Leslie B. Davis and John W. Fisher Jr. (Salt Lake City: University of Utah Press, 2016), 150.

CHAPTER 8

1. Eileen Johnson, "Lubbock Lake Landmark – A Joint Venture in Preservation, Research, and Public Interpretation," in *Interpreting First Peoples Bison Kills at Heritage Parks*, ed. Leslie B. Davis and John W. Fisher Jr. (Salt Lake City: University of Utah Press, 2016), 129.
2. Johnson, "Lubbock Lake Landmark – A Joint Venture," 153.
3. Barbara Whitney, "Sue, Dinosaur Skeleton," *Britannica*, updated October 18, 2024, https://www.britannica.com/topic/Sue.
4. "Accreditation and Excellence Programs," American Alliance of Museums, https://www.aam-us.org/programs/accreditation-excellence-programs/.
5. Institute of Museum and Library Sciences, "Government Doubles Official Estimates: There are 35,000 Active Museums in the U.S.," Washington, DC, May 19, 2–14, https://www.imls.gov/news/government-doubles-official-estimate-there-are-35000-active-museums-us.
6. Erin Scheaffer, contributor, "Monuments Men: Preserving Cultural Heritage During a Period of Great Turmoil," The National World War II Museum, New Orleans, May 28, 2020, https://www.nationalww2museum.org/war/articles/monuments-men.
7. "The 1954 Hague Convention and Its Protocols," UNESCO, 2020, https://www.unesco.org/archives/multimedia/document-4983.
8. "Horn Distinguished Professors," Office of the Provost, Texas Tech University, https://www.depts.ttu.edu/provost/

horn-professors/.

9. Susan Leigh Tomlinson, "How to Keep a Naturalist's Notebook" (Mechanicsburg, PA: Stackpole Books, 2010).
10. Johnson, "Lubbock Lake Landmark – A Joint Venture," 129–66.

CHAPTER 9

1. "Why Study History," University of Illinois Department of History, https://history.illinois.edu/student-financial-aid/why-study-history.
2. Frances Mayhugh Holden, "Holden, William Curry," *Handbook of Texas Online*, Texas State Historical Association, https://www.tshaonline.org/handbook/entries/holden-william-curry.
3. John T. Montford and Joseph Daniel McCool, *Board Games: Straight Talk for New Directors and Good Governance* (n.p.: Praeger, June 30, 2016), https://boardgamesbook.com/authors/john-t-montford/.
4. "Horn Distinguished Professors," Office of the Provost, Texas Tech University, https://www.depts.ttu.edu/provost/horn-professors/.
5. Vance T. Holliday, "Eileen Johnson's Legacy of Research," *Quaternary International* 492 (October 2018): 2–9, https://www.sciencedirect.com/journal/quaternary-international/vol/492/suppl/C.
6. William Shakespeare, *The Tempest*, Act 2, Scene 1, The Folger Shakespeare Library, last updated 2024, https://www.folger.edu/explore/shakespeares-works/the-tempest/.

BIBLIOGRAPHY

"About Wildfires." Smokey Bear, 2021. https://smokeybear.com/.

"Accreditation and Excellence Programs." American Alliance of Museums. https://www.aam-us.org/programs/accreditation-excellence-programs/.

"American Lions." National Park Service, January 27, 2002. https://www.nps.gov/whsa/learn/nature/american-lions.htm.

American Museum of Natural History. *The Evolution of Horses*. American Museum of Natural History Part of the Horse exhibition, May 17, 2008–January 4, 2009. https://www.amnh.org/exhibitions/horse/the-evolution-of-horses.

"Ancient Bison." National Park Service. https://www.nps.gov/articles/000/ancient-bison.htm.

"Ancient Oasis on the Southern Plains." Volunteer material from Lubbock Lake Landmark.

Anderson, Hugh Allen. "Black Horse [TuUkumah]." *Handbook of Texas Online*. Texas State Historical Association, accessed June 9, 2025. https://www.tshaonline.org/handbook/entries/black-horse-tu-ukumah.

———. "Pastores." *Handbook of Texas Online*. Texas State Historical Association, May 3, 2019. https://www.tshaonline.org/handbook/entries/pastores.

———. "Singer, George Washington." *Handbook of Texas Online*. Texas State Historical Association, updated February 3, 2021. https://www.tshaonline.org/handbook/entries/singer-george-washington.

———. "Yellow House Canyon, Battle of." *Handbook of Texas Online*. Texas State Historical Association, April 7, 2021. https://www.tshaonline.org/handbook/entries/yellow-house-canyon-battle-of.

"Archaeologist Demographics in the United States." Career Explorer, 2024. https://www.careerexplorer.com/careers/archaeologist/demographics/.

"Archaic Period." National Park Service. https://www.nps.gov/articles/archaic-period.htm.

Bigness, Deborah. Interview by Sally Logue Post, December 13, 2022. Lubbock, Texas.

"Bison, Buffalo, Tatanka: Bovids of the Badlands." Wildlife in the Badlands, National Parks Service. https://www.nps.gov/articles/bison_badl.htm.

Black, Craig, ed. *History and Prehistory of the Lubbock Lake Landmark*. Lubbock: West Texas Museum Association, 1974.

Blackwater Draw National Historic Landmark and Museum. Eastern New Mexico University, 2017–2026. https://www.bwdarchaeology.com/bwdnationalhistoriclandmark.

Blair, Keith. Interview by Sally Logue Post, December 14, 2022. Via Zoom.

Brazos River Authority. "The Brazos River." https://brazos.org/about-us/about-the-bra/about-the-brazos-river.

Britannica editors. "Folsom Complex." *Encyclopedia Britannica*, January 5, 2012. https://www.britannica.com/topic/Folsom-complex.

Brusatte, Steve. *The Rise and Reign of the Mammals: A New History, from the Shadow of the Dinosaurs to Us*. New York: Mariner Books, 2023.

Butler, Kristina Woods. "Preserving History for the Future." *Discoveries*, Spring 2012. Texas Tech University.

"Camel – Camel History." Better Planet Education. https://ypte.org.uk/factsheets/camel/camel-history?hide_donation_prompt=1.

"The Camels." National Park Service. https://www.nps.gov/articles/000/the-camels.htm.

"City to Take First Step in Lake Site Development." *Lubbock Avalanche-Journal*, July 28, 1966.

"Clovis Complex." *Encyclopedia Britannica*, September 21, 2024. https://www.britannica.com/topic/Clovis-complex.

"Columbian Mammoth: Tule Springs Fossil National Monument." National Park Service, updated January 31, 2024. https://www.nps.gov/articles/000/columbian-mammoth.htm.

"Confederated Salish and Kootenai Tribes." Office of the Governor of Montana, Indian Affairs. https://tribalnations.mt.gov/Directory/ConfederatedSalishKootenaiTribes.

"Cosmic Origins: Giving the Clock Its Tick." Lubbock Lake Landmark Nash Interpretive Center. Lubbock, Texas, permanent exhibit, opened 1990.

Crawford, Markus. Interview by Sally Logue Post, May 31, 2023. Stanton, Texas.

Cruse, J. Brett. *Red River War: Archeological Perspectives on the Indian Campaign of 1874*. Austin: Texas Historical Commission, 2017, 2nd printing.

"Cultural Sequence." Lubbock Lake Landmark. https://www.depts.ttu.edu/museumttu/lll/Context.html.

Cunningham, Doug. Interviews by Sally Logue Post, November 30, 2022, and December 23, 2024. Lubbock, Texas.

Cutrer, Thomas W. "Marcy, Randolph Barnes (1812–1887)." *Handbook of Texas Online*. Texas State Historical Association, December 9, 2020. https://www.tshaonline.org/handbook/entries/marcy-randolph-barnes.

Dalquest, Walter W. "Vertebrae Fossils from the Blanco Local Fauna of Texas." *Occasional Papers. The Museum of Texas Tech University*, no. 30 (May 23, 1975): 1–2.

Davis, Charles G. "Yellowhouse Draw." *Handbook of Texas Online*. Texas State Historical Association, updated February 1, 1996. https://www.tshaonline.org/handbook/entries/yellow-house-draw.

DeJesse, Andrew. Interview by Sally Logue Post, June 20, 2023. Lubbock, Texas.

Deposition Processes. Lubbock Lake Landmark Nash Interpretive Center, Lubbock, Texas. Permanent exhibition, opened 1990 to present.

Ditore, Cameron. Interview by Sally Logue Post, May 31, 2023. Stanton, Texas.

Donovan, Geoffrey H., and Thomas C. Brown. "Be Careful What You Wish For: The Legacy of Smokey Bear." *Frontiers in Ecology and the Environment* 5, no. 2 (2007): 73–79.

Ehlers, Katherine. Interview by Sally Logue Post, January 25, 2023. Via Zoom.

Engaging Folsom (10,800–10,200) Hunter-Gatherers with 3D Technologies. Lubbock Lake Landmark Nash Interpretive Center exhibit, on display 2017. Lubbock, Texas.

Erichsen, Gerald. "The Definition of Llano Estacado in the U.S. and Around the World." ThoughtCo, June 3, 2019. https://www.thoughtco.com/what-does-llano-estacado-mean-3971910.

"Ethnobotany of Mesquite Trees." National Park Service, July 4, 2020. https://www.nps.gov/articles/000/ethnobotany-of-mesquite-trees.htm.

Evans, Glen L., and Grayson E. Meade. *Quaternary of the Texas High Plains*. University of Texas, Bureau of Economic Geology, publication 4401 (1945): 485–507.

Farmer, Jesse R., Tamara Pico, Ona M. Underwood, and Daniel M. Sigman. "The Bering Strait Was Flooded 10,000 Years Before the Last Glacial Maximum." *Proceedings of the National Academy of Sciences* 119, no. 52 (December 27, 2022). https://www.pnas.org/doi/10.1073/pnas.2206742119.

"Field Guide for Managing Siberian Elm in the Southwest." United States Department of Agriculture, September 2014. https://www.fs.usda.gov/Internet/FSE_DOCUMENTS/stelprdb5410128.pdf.

"Fire on the Land: Native People and Fire in the Northern Rockies." Confederated Salish and Kootenai Tribes – Division of Fish, Wildlife, Recreation, and Conservation, 2021. http://fwrconline.csktnrd.org/Fire/FireOnTheLand/.

"Forest Resources." *Texas Almanac*, 2022. https://www.texasalmanac.com/articles/forest-resources.

Fossils Through Geologic Time. National Park Service, updated December 11, 2024. https://www.nps.gov/subjects/fossils/fossils-through-geologic-time.htm.

Foster, Jerod. Interview by Sally Logue Post, July 7, 2023. Lubbock, Texas.

Gard, Wayne. *The Great Buffalo Hunt.* Alfred A. Knopf, 1959.

"Geology of Rocky Mountain National Park." Geology and Ecology of National Parks. United States Geological Survey. https://www.usgs.gov/geology-and-ecology-of-national-parks/geology-rocky-mountain-national-park.

"The Giant Short-Faced Bear." North American Bear Center. https://bear.org/bear-facts/the-giant-short-faced-bear/.

Goodman, George J., and Cheryl A. Lawson. *Retracing Major Stephen H. Long's 1820 Expedition*. Norman: University of Oklahoma Press, 1995.

Goss, Jim. Interview by Daniel U. Sánchez, October 8, 2014. Oral history online transcription. Southwest Collection/Special Collections Library, Texas Tech University. https://swco-ir.tdl.org/server/api/core/bitstreams/8b269ad9-001e-4abc-8854-c15657e0e9e7/content.

"Government Doubles Official Estimate: There Are 35,000 Active Museums in the U.S." Washington, DC: Institute of Museum and Library Services, May 19, 2024. https://www.imls.gov/news/government-doubles-official-estimate-there-are-35000-active-museums-us.

Graves, Lawrence L., ed. *A History of Lubbock*. Lubbock: Texas Technological College, West Texas Museum Association, 1962.

Greiner, Alyson L. "Dugouts." *Encyclopedia of Oklahoma History and Culture*. https://www.okhistory.org/publications/enc/entry?entry=DU003.

Haley, James L. *The Buffalo War: The History of the Red River Indian Uprising of 1874*. Austin: State House Press, 1998.

———. "Red River War." *Handbook of Texas Online*. Texas State Historical Association, updated January 27, 2021. https://www.tshaonline.org/handbook/entries/red-river-war#:~:text=Frank%20D.,the%20ranching%20era%20followed%20swiftly.

Hamilton, Lucchese. Interview by Sally Logue Post, May 18, 2023. Lubbock, Texas.

"Henry A. Wright Memorial Endowed Scholarship." Davis College of Agricultural Sciences and Natural Resources, Texas Tech University, September 28, 2023. https://www.depts.ttu.edu/agriculturalsciences/Students/scholarships/nrm/wright.php.

Hickey, Hannah. *What Caused Earth's Biggest Mass Extinction?* Doerr School of Sustainability, Stanford University, December 6, 2018. https://sustainability.stanford.edu/news/what-caused-earths-biggest-mass-extinction.

Hicks, J. Kent, and Eileen Johnson. "Pastores Presence on the Southern High Plains of Texas." *Historical Archaeology* 34, no. 4 (2000): 46–60. http://www.jstor.org/stable/25616851.

"Historic Maintenance." *Notes from the Field* (Lubbock Lake Landmark newsletter), Fall 2002.

Holden, Frances Mayhugh. "Holden, William Curry." *Handbook of Texas Online*. Texas State Historical Association. https://www.

tshaonline.org/handbook/entries/holden-william-curry.

Holden, W. C. "Historical Background of the Lubbock Lake Site." *The Museum Journal* 15 (1974): 13.

———. "The Land." *A History of Lubbock*, Part 1, edited by Lawrence L. Graves. West Texas Museum Association, 1959.

———. "U Lazy S Ranch." *Handbook of Texas Online*. Texas State Historical Association, January 23, 2019. https://www.tshaonline.org/handbook/entries/u-lazy-s-ranch.

Holliday, Vance T. "Cultural Chronology of the Lubbock Lake Site." Master's thesis, Texas Tech University, Department of Museum Science, 1977. https://ttu-ir.tdl.org/server/api/core/bitstreams/6268c501-ec0b-49de-be59-2372f3ca8de4/content.

———. "Eileen Johnson's Legacy of Research." *Quaternary International* 492 (October 2018), 2–9. https://www.sciencedirect.com/journal/quaternary-international/vol/492/suppl/C.

———. "Guidebook to the Central Llano Estacado." *Friends of the Pleistocene South-Central Cell Field Trip* (ICASALS and the Museum of Texas Tech University).

———. Interviews by Sally Logue Post, October 18, 2022, and July 3, 2024. Via Zoom.

Holliday, Vance T., Susan D. Hovorka, and Thomas C. Gustavson. "Lithostratigraphy and Geochronology of Fills in Small Playa Basins on the Southern High Plains." *Geological Society of America Bulletin* 108 (1996): 953–65.

Holliday, Vance T., Jason D. Windingstad, Jordon Bright, Bruce G. Phillips, Joel B. Butler, Ryan Breslawski, and James E. Bowman. "Paleolake Geochronology Supports Last Glacial Maximum (LGM) Age for Human Tracks at White Sands, New Mexico." *Science Advances* 11, no. 25 (2023): eadh5006. https://www.science.org/doi/10.1126/sciadv.adv4951.

Holt, Katie. Interview by Sally Logue Post, March 27, 2023. Lubbock, Texas.

Hopper, Kippra D. Interview by Sally Logue Post, January 9, 2023.

Lubbock, Texas.

———. *Visionary, William Curry Holden*. Unpublished manuscript, 2007.

"Horn Distinguished Professors." Office of the Provost, Texas Tech University. https://www.depts.ttu.edu/provost/horn-professors/.

Hosmer, Brian C. "Parker, Quanah." *Handbook of Texas Online*. Texas State Historical Association, updated January 7, 2021. https://www.tshaonline.org/handbook/entries/parker-quanah.

Hurst, Stance. "Dating the Whiskey Flats Site with a Glass Bead and Bison Ear Bone." *Notes from the Field* (Lubbock Lake Landmark newsletter), Fall 2016.

———. Interviews by Sally Logue Post, January 9, 2023; January 13, 2023; February 15, 2023; March 1, 2023; May 31, 2023; December 15, 2023; and October 27, 2024. Lubbock, Texas.

———. "Licking Horse Bones in Mustang Draw." *Notes from the Field* (Lubbock Lake Landmark newsletter), Winter 2013.

———. "A New Clovis Paleoindian Site Discovery at Macy Locality 10." *Notes from the Field* (Lubbock Lake Landmark Newsletter), Spring 2015.

———. "Recent Work at Tahoka Lake." *Notes from the Field* (Lubbock Lake Landmark newsletter), Fall 2016.

Hurst, Stance, and Eileen Johnson. "Adair-Steadman (41FS2) – Survey at a Folsom Site in the Brazos River Drainage on the Southern Plains." *Index of Texas Archaeology* 1, no. 26 (January 2019). https://scholarworks.sfasu.edu/ita/vol2019/iss1/26/.

———. "Gearing Up at the Adair-Steadman (41FS2)." *PaleoAmerica* 2, no. 3 (August 18, 2016): 252–60. https://www.tandfonline.com/doi/citedby/10.1080/20555563.2016.1198215?scroll=top&needAccess=true.

———. "Prominent Places on the Landscape: Occupations at Cowhead Mesa Along the Eastern Escarpment Edge of the Southern High Plains of Texas." *Plains Anthropologist* 62, no. 241 (2016): 5–31. https://www.tandfonline.com/doi/full/10.1

080/00320447.2016.1235303.

Hurst, Stance, Elizabeth Louden, and Eileen Johnson. "Unravelling the Cowhead Mesa Petroglyphs with 3-D Laser Scanning and High-Resolution Digital Photography." *Rock Art Research* 26, no. 1 (May 2009): 43–53. https://www.researchgate.net/publication/265085921_Unravelling_the_Cowhead_Mesa_Petroglyphs_with_3-D_Laser_Scanning_and_High-Resolution_Digital_Photography.

Hurst, Stance, Dallas C. Ward, Eileen Johnson, and Doug Cunningham. "Cowboy Life Along the Llano Estacado Eastern Escarpment of Northwest Texas: Insights from Macy Locality 16." *Historical Archaeology* 52 (March 30, 2018): 332–47.

Institute of Museum and Library Sciences. "Government Doubles Official Estimates: There are 35,000 Active Museums in the U.S." Washington, DC, May 19, 2–14. https://www.imls.gov/news/government-doubles-official-estimate-there-are-35000-active-museums-us. "Interpretive Signs." Lubbock Lake Landmark Trails and Sculptures.

Jenkins, John Cooper. *Estacado: Cradle of Culture and Civilization on the Staked Plains of Texas*. Crosby County Pioneer Memorial Museum, 1986.

Johnson, Eileen. Interviews by Sally Logue Post, September 1, 2022; September 21, 2022; October 5, 2022; October 19, 2022; November 10, 2022; November 21, 2022; December 15, 2022; January 18, 2023; February 16, 2023; October 19, 2023; October 25, 2023; September 25, 2024; and December 15, 2024. Lubbock, Texas.

———. "Landscapes and Peoples of the Llano Estacado." In *Archaeological Landscapes on the High Plains*, edited by Laura L. Scheiber and Bonnie J. Clark, 115–56. Denver: University Press of Colorado, 2008.

———. "Lubbock Lake Landmark – A Joint Venture in Preservation, Research, and Public Interpretation." In *Interpreting First*

Peoples Bison Kills at Heritage Parks, edited by Leslie B. Davis and John W. Fisher Jr., 129–66. Salt Lake City: University of Utah Press, 2016.

Johnson, Eileen, and Vance T. Holliday. "Archaeology and Late Quaternary Environments of the Southern High Plains." In *The Prehistory of Texas*, edited by Timothy K. Perttula. College Station: Texas A&M University Press, 2012. Ebook, accessed via Kindle.

———. "The Archaic Record at Lubbock Lake." *Plains Anthropologist* 31, no. 114 (November 1986): 7–54. https://www.jstor.org/stable/25668646.

———. "Introduction." In *Lubbock Lake Late Quaternary Studies on the Southern High Plains*, edited by Eileen Johnson. College Station: Texas A&M University Press, 1987.

Johnson, Eileen, and John A. Moretti. "Diversity and Paleoecology of the Zone 3 Late Pleistocene Vertebrates at Kincaid Shelter (41UV2), Central Texas, USA." *PaleoAmerica: A Journal of Early Human Migration and Dispersal* 8, no. 3 (June 26, 2022): 228–52. https://www.tandfonline.com/doi/abs/10.1080/20555563.2022.2075134.

Johnson, Eileen, and Susan Shore. *Mammals at Lubbock Lake Landmark*. Lubbock: Lubbock Lake Landmark Publication, 2010.

Jones, Lila. Interviews by Sally Logue Post, March 27, 2023. Lubbock, Texas; May 31, 2023. Stanton, Texas.

Judson, Olivia. "When Texas Was at the Bottom of the Sea." *Smithsonian* magazine, January 2015. https://www.smithsonianmag.com/travel/when-texas-was-bottom-sea-180953653/#:~:text=The%20Permian%20Period%20ran%20from,most%20catastrophic%20extinction%20on%20record.

Judson, Sheldon. "Geology of the San Jon Site, Eastern New Mexico." *Smithsonian Miscellaneous Collections* 121, no. 1, March 5, 1953. https://repository.si.edu/bitstream/handle/10088/22882/SMC_121_Judson_1953_1_1-70.pdf.

Kelley, Jane Holden. "A Brief Resume of Artifacts Collected at the Lubbock Lake Site Prior to 1961." *The Museum Journal* 15 (1974): 43–78.

Klotzman, Jeff. Interview by Sally Logue Post, January 9, 2024. Via Zoom.

Lane, Wendy E. "Experts: Lubbock May Be One of the New World's Oldest Towns." *Dallas Morning News*, December 26, 1988.

Leatherwood, Art. "Llano Estacado." *Handbook of Texas Online*. Texas State Historical Association, June 18, 2020. https://www.tshaonline.org/handbook/entries/llano-estacado.

Lehmann, Herman. *Nine Years Among the Indians, 1870–1879*. N.p.: Arcadia Press, 2019.

Library of Congress. "The American West, 1865–1900." U.S. History Primary Source Timeline, accessed July 23, 2025. https://www.loc.gov/classroom-materials/united-states-history-primary-source-timeline/rise-of-industrial-america-1876-1900/american-west-1865-1900/.

"Life in a Dugout Hunting Buffalo and Herding Cattle." Lubbock Lake Landmark Nash Interpretive Center, temporary exhibit. Lubbock, Texas, 2023.

Lipscomb, Carol A. "Comanche Indians." *Handbook of Texas Online*. Texas State Historical Association, October 9, 2020. https://www.tshaonline.org/handbook/entries/comanche-indians.

Lubbock City Council. Recognize and Commend Robert A. "Bob" Nash. R1988-2719, January 28, 1988.

Lubbock Lake Landmark. *Lubbock Lake Landmark Master Plan: Quaternary Research and Archaeology Education Facility*, vols. 1, 2, and 3. Lubbock, Texas, December 1983.

"Lubbock Lake Landmark Natural History Series." *Occasional Paper: The Museum of Texas Tech University* no. 30 (May 23, 1975): 1–2.

Macy, Bob. Interviews by Sally Logue Post, January 24, 2023, via phone; February 21, 2023, Lubbock, Texas.

"Mammals at Lubbock Lake Landmark." *Natural History Series.* A Lubbock Lake Landmark publication, 2010.

"Mammals Long Ago." Volunteer material from Lubbock Lake Landmark.

Mann, Charles C. "The Clovis Point and the Discovery of America's First Culture." *Smithsonian*, November 2013. https://www.smithsonianmag.com/history/the-clovis-point-and-the-discovery-of-americas-first-culture-3825828/.

"Maps." The King Ranch. https://king-ranch.com/about-us/maps/.

"Marcy, Randolph Barnes (1812–1887)." *Encyclopedia of Oklahoma History and Culture.* Oklahoma Historical Society. https://www.okhistory.org/publications/enc/entry?entry=MA021.

Matthew, Glenn Hill, Vance T. Holliday, and Dennis J. Stanford. "A Further Evolution of the San Jon Site, New Mexico." *Plains Anthropologist* 40, no. 154 (November 1995): 369–90. https://www.jstor.org/stable/25669365.

May, Clyde. Interview by Sally Logue Post, September 9, 2023. Tahoka Lake Pasture Preserve.

McCarty, Jerry. "City Gives Boost to Lake Site Park." *Lubbock Avalanche-Journal*, May 11, 1967.

———. "Lack of Funds May Delay Lubbock Lake Site Park." *Lubbock Avalanche-Journal*, October 13, 1967.

———. "Park Development Plans Aired Here." *Lubbock Avalanche-Journal*, April 5, 1968.

McIntire, Shannon. "Can You Dig It?" Texas Department of Transportation Newsroom, May 13, 2025. https://www.txdot.gov/about/newsroom/stories/can-you-dig-it.html.

McKelvey, Daria. "The Beauty of the Lubbock Lake Landmark." *Notes from the Field* (Lubbock Lake Landmark newsletter), Summer 2020.

———. Interview by Sally Logue Post. January 22, 2023. Via Zoom.

Mendoza, Graciela. Interview by Sally Logue Post, April 3, 2024. Lubbock, Texas.

Mendoza, José. Interview by Sally Logue Post, April 3, 2024. Lubbock, Texas.

"Mexican on the Rampage." *Galveston Daily News*, April 26, 1886.

"Monte Verde Archaeological Site." UNESCO World Heritage Convention, submitted February 23, 2004. https://whc.unesco.org/en/tentativelists/1873/.

Montford, John T. Interview by Sally Logue Post, July 19, 2023. Lubbock, Texas.

Montford, John T., and Joseph Daniel McCool. *Board Games: Straight Talk for New Directors and Good Governance*. N.p.: Praeger, June 30, 2016. https://boardgamesbook.com/authors/john-t-montford/.

Moretti, John. "Ancient Deposits Reveal Extinct and Extant Animals of the Southern High Plains." *Notes from the Field* (Lubbock Lake Landmark newsletter), Winter 2013.

———. "Giant Tortoise Adds to the Exotic Animals of Macy Locality 100." *Notes from the Field* (Lubbock Lake Landmark newsletter), Spring 2014.

———. "The Great Camel Discovery." *Notes from the Field* (Lubbock Lake Landmark newsletter), Summer 2009.

———. "The Ice Age Lion of the Southern High Plains." *Notes from the Field* (Lubbock Lake Landmark newsletter), Spring 2017.

———. Interviews by Sally Logue Post, February 21, 2023, via Zoom; July 20, 2023, Lubbock, Texas; March 31, 2024, and May 31, 2024, via Zoom.

———. "Pleistocene Proboscidean Tusk Segments Discovered." *Notes from the Field* (Lubbock Lake Landmark newsletter), Winter 2013.

———. "Precarious Excavation Produces a Wealth of Ice Age Animals." *Notes from the Field* (Lubbock Lake Landmark newsletter), Summer 2020.

Morris, Blake. "Tumble Weeds." *Notes from the Field* (Lubbock Lake Landmark newsletter), Fall 2007.

"A Most Misunderstood Landscape." Lubbock Lake Landmark Nash Interpretive Center. November 2017–October 2018.

Mullican, Shea. Interview by Sally Logue Post, October 17, 2024. Lubbock, Texas.

Myers, Mindy, Eileen Johnson, Susan Shore, Vance T. Holliday, and Susan Rowe. *The Lubbock Lake Landmark.* Lubbock Lake Natural History Series. Lubbock: Lubbock Lake Landmark Publication, 2012.

Neelley, Bert. "C. of C. Support Given Lake Site." *Lubbock Avalanche-Journal*, July 13, 1966.

"The 1954 Hague Convention and its Protocols." UNESCO, 2020. https://www.unesco.org/archives/multimedia/document-4983.

Ogallala Aquifer Initiative 2011 Report. Natural Resources Conservation Service, United States Department of Agriculture. https://cybercemetery.unt.edu/archive/oilspill/20130225074112/http://www.ndcsmc.nrcs.usda.gov/Internet/FSE_DOCUMENTS/stelprdb1048827.pdf.

Osterloff, Emily. "Tumbleweeds: The Fastest Plant Invasion in the USA's History." Natural History Museum, London, accessed May 18, 2024. https://www.nhm.ac.uk/discover/tumbleweeds-fastest-plant-invasion-in-usa-history.html.

Pan, Aaron. Interview by Sally Logue Post, February 6, 2024. Lubbock, Texas.

Parrish, Curtis. Interview by Sally Logue Post, February 9, 2023. Lubbock, Texas.

Paselk, Richard. *Permian Period.* Cal Poly Humboldt Natural History Museum, revised October 30, 2012. https://natmus.humboldt.edu/exhibits/life-through-time/visual-timeline/permian-period.

Peoples, Matt. "George McJunkin and the Discovery That Changed

American Archaeology." *Archaeology Southwest* (blog), February 23, 2015. https://www.archaeologysouthwest.org/2015/02/23/george-mcjunkin-and-the-discovery-that-changed-american-archaeology/.

Perkins, Sid. "Ancient Armadillos Grew as Big as VW Beetles." *Science*, February 22, 2016. https://www.science.org/content/article/ancient-armadillos-grew-big-vw-beetles.

Pettit, Keller. Interview by Sally Logue Post, April 5, 2024. Lubbock, Texas.

Pinson, Jerald. "Scientists Have Figured Out How Extinct Giant Ground Sloths Got So Big—and Where It All Went Wrong." Phys.org. Florida Museum of Natural History, May 22, 2025. https://phys.org/news/2025-05-scientists-figured-extinct-giant-ground.html.

Pitts, Joseph. Interview by Sally Logue Post, October 17, 2024. Lubbock, Texas.

"Plant Data Base: Tahoka Daisy." Lady Bird Johnson Wildflower Center, January 01, 2010. https://www.wildflower.org/plants/result.php?id_plant=MATA2.

"Prescribed Burning." Texas Parks and Wildlife. https://tpwd.texas.gov/landwater/land/habitats/post_oak/habitat_management/fire/.

"Prescribed Burning Board." Texas Department of Agriculture. https://texasagriculture.gov/Home/Production-Agriculture/Prescribed-Burn-Program/Prescribed-Burning-Board.

"Quanah Parker." The Historical Marker Database, December 21, 2017. https://www.hmdb.org/m.asp?m=111756.

"Quaternary Period." *National Geographic*. https://www.nationalgeographic.com/science/article/quaternary.

Roos, Dave. "How Early Humans First Reached the Americas: 3 Theories." *History*, last updated March 17, 2025. https://www.history.com/articles/human-migration-americas-beringia.

Rowe, Susan. Interviews by Sally Logue Post, December 15, 2022,

February 18, 2023, and February 28, 2023. Lubbock, Texas.

Sanders, Robert. "Tests Confirm Humans Tramped Around North America More Than 20,000 Years Ago." *Berkeley News*, October 5, 2023. https://news.berkeley.edu/2023/10/05/tests-confirm-humans-tramped-around-north-america-more-than-20-000-years-ago.

Sansom, Andrew. Interview by Sally Logue Post, December 27, 2022. Via Zoom.

Scheaffer, Erin, contributor. "Monuments Men: Preserving Cultural Heritage During a Period of Great Turmoil." The National World War II Museum, New Orleans, May 28, 2020. https://www.nationalww2museum.org/war/articles/monuments-men.

Shakespeare, William. *The Tempest*. The Folger Shakespeare Library, last updated 2024. https://www.folger.edu/explore/shakespeares-works/the-tempest/.

Shelton, Sally. Interview by Sally Logue Post, June 20, 2023. Lubbock, Texas.

Simpson, Cullom, Whitney Gann, Bonnie Warnock, and Louis A. Harveson. "Project Spotlight: Prairie Dogs and Cattle." Borderlands Research Institute, Sul Ross State University, May 27, 2021. https://bri.sulross.edu/project-spotlight-prairie-dogs-and-cattle/#.

"Singer Store and Family." Volunteer material. Lubbock Lake Landmark.

"Smokey Bear About Wildfires." Ad Council. https://smokeybear.com/.

Society for American Archaeology. "What Do Archaeologists Do?" https://saa.org/Member/SAAMember/About-Archaeology/What-do-Archaeologists-do.aspx.

Swenson, Kyle. "Was Smokey Bear Wrong? How a Beloved Character May Have Helped Fuel Catastrophic Fires." *Washington Post*, August 15, 2018.

Taylor, Anna Jean. *Pastores in the Texas Panhandle, 1876–1884.*

Thesis, Wayland Baptist College, Spring 1976.

Texan by Nature. "A Conservation Guide for Texas Landowners." https://texanbynature.org /2023/03/a-conservation-guide-for-texas-landowners/.

Texas Historical Commission. "Red River War of 1874–1875: Clash of Cultures in the Texas Panhandle, A Texas Travel Guide." https://www.thc.texas.gov/public/upload/publications/red-river-war.pdf.

"Texas Horned Lizard (*Phrynosoma cornutum*)." Texas Parks and Wildlife. https://tpwd.texas.gov/huntwild/wild/species/thlizard/.

Texas Parks and Wildlife. *Lubbock Lake Landmark State Historic Site, Master Plan and Site Analysis.* Austin, Texas, January 1988.

Tomlinson, Susan Leigh. *How to Keep a Naturalist's Notebook.* Mechanicsburg, PA: Stackpole Books, 2010.

———. Interview by Sally Logue Post, March 7, 2023. Lubbock, Texas.

Trevey, Sterling Scott. Interview by Sally Logue Post, March 11, 2024. Lubbock, Texas.

———. "Landscape Changes Create New Landmark Views." *Notes from the Field* (Lubbock Lake Landmark newsletter, Spring 2005.

———. "Native, Notable, and Noxious." *Notes from the Field* (Lubbock Lake Landmark newsletter), Fall 2007.

———. "Old and New." *Notes from the Field* (Lubbock Lake Landmark newsletter), Spring 2004.

Vigness, David. "The History of Buffalo Hunting on the Great Plains." *Handbook of Texas Online.* Texas State Historical Association, last updated June 1, 1995. https://www.tshaonline.org/handbook/entries/buffalo-hunting.

Wade, Sarah. "8 Surprising Prairie Dog Facts." World Wildlife Fund. https://www.worldwildlife.org/resources/explainers/prairie-dogs-the-tiny-architects-of-the-great-plains/.

Ward, Dallas. "A Heritage Perspective on the Historic U-Lazy-S Ranch – A Case Study in Heritage Construction." Master's thesis, Texas Tech University, Department of Heritage and Museum Science, August 2016. https://hdl.handle.net/2346/82260.

"Water, Man, and Lubbock Lake." Lubbock Lake Landmark Nash Interpretive Center. Lubbock, Texas.

Wheat, Joe Ben. "First Excavations at the Lubbock Lake Site." *The Museum Journal* 15 (1974): 15.

Whitney, Barbara. "Sue, Dinosaur Skeleton." *Britannica*, updated October 18, 2024. https://www.britannica.com/topic/Sue.

"Whose Buffalo? - Fact Sheet." Texas Beyond History. https://www.texasbeyondhistory.net/redriver/images/rr-whosebuffalo.pdf.

"Why Study History." University of Illinois Department of History. https://history.illinois.edu/student-financial-aid/why-study-history.

"Wildfire Causes." Western Fire Chiefs Association, March 4, 2024. https://wfca.com/wildfire-articles/what-causes-wildfires/.

"Wildland Fire Program." National Park Service, January 12, 2022. https://www.nps.gov/orgs/1965/wildfires-prescribed-fires-fuels.htm.

Wiley, Taylor, Valeria Chamorro, Joshua J. Wells, Eric Kansa, and Patrick Finnegan. "More Than Just Another Number: Use of the Smithsonian Trinomial System and the Digital Index of North American Archaeology (DINAA) to Link Open Information about Archaeological Sites Across the Web." Society for American Archaeology 82nd Annual Meeting. Vancouver, BC, March 2017.

Wilkes, Morris. Interview by Sally Logue Post, October 18, 2022. Lubbock, Texas.

Wilson, C. Dean. "Chupadero Black-on-White." Office of Archaeological Studies Pottery Typology Project, 2012. https://ceramics.nmarchaeology.org/typology/type?p=287.

Woody, J. W. Letter to J. Evetts Haley, October 10, 1926. J. Evetts Haley Collection, Haley Memorial Library and History Center. Midland, Texas.

———. Letter to J. Evetts Haley. February 11, 1928. J. Evetts Haley Collection, Haley Memorial Library and History Center. Midland, Texas.

Yater, Lyn. Interview by Sally Logue Post, May 31, 2023. Stanton, Texas.

Zartman, Richard. Interview by Sally Logue Post, November 11, 2022. Lubbock, Texas.

Zimmerman, Kim Ann. "Cenozoic Era: Facts About Climate, Animals and Plants." Live Science, June 8, 2016. https://www.livescience.com/40352-cenozoic-era.html.

INDEX

Note: Page numbers in italics refer to figures.

ABOUT THE AUTHOR

Sally Logue Post spent twenty-five years at Texas Tech University in communications and marketing roles for the university, the Office of the Vice President for Research, and the Museum of Texas Tech University. She also worked as a journalist for two decades at Texas newspapers and television stations. She holds bachelor's and master's degrees from Texas Tech. She is now retired.

www.ingramcontent.com/pod-product-compliance
Lightning Source LLC
La Vergne TN
LVHW091535010926
846999LV00030B/279